Samia ARAAR
Fadila Khaldi
Sarra SAYAH

Fertilizer toxicity in rats

AF665900

Samia ARAAR
Fadila Khaldi
Sarra SAYAH

Fertilizer toxicity in rats

Evaluation of the toxic effects of calcium nitrate fertilizer on white rats

ScienciaScripts

Imprint
Any brand names and product names mentioned in this book are subject to trademark, brand or patent protection and are trademarks or registered trademarks of their respective holders. The use of brand names, product names, common names, trade names, product descriptions etc. even without a particular marking in this work is in no way to be construed to mean that such names may be regarded as unrestricted in respect of trademark and brand protection legislation and could thus be used by anyone.

Cover image: www.ingimage.com

This book is a translation from the original published under ISBN 978-620-6-72609-8.

Publisher:
Sciencia Scripts
is a trademark of
Dodo Books Indian Ocean Ltd. and OmniScriptum S.R.L publishing group

120 High Road, East Finchley, London, N2 9ED, United Kingdom
Str. Armeneasca 28/1, office 1, Chisinau MD-2012, Republic of Moldova, Europe
Managing Directors: Ieva Konstantinova, Victoria Ursu
info@omniscriptum.com

Printed at: see last page
ISBN: 978-620-8-52147-9

Copyright © Samia ARAAR, Fadila Khaldi, Sarra SAYAH
Copyright © 2024 Dodo Books Indian Ocean Ltd. and OmniScriptum S.R.L publishing group

Table of contents

INTRODUCTION GENERAL

Global food demand will continue to grow over the coming decades with concomitant challenges for sustainable agricultural supply (Conijn et *al.*, 2018; Gouel & Guimbard, 2019). The immediate response to this growing demand for food production is intensive use of agrochemicals, which largely include fertilizers and pesticides (Carvalho, 2006). In addition, intensive agriculture, geared to maximum productivity and yield, has developed considerably and has also developed the massive use inputs (fertilizers and pesticides) (Shah & Wu, 2019).

Fertilizers are all fertilizing materials dispersed on the soil surface and intended to provide plants with the elements they need to build up their tissues (Testud, 2004). They are mixtures of chemicals containing one or more of the three basic nutrients (nitrogen, phosphorus and potassium) needed to stimulate plant growth and agricultural production (EFMA 1997). They are also a source of secondary nutrients, including sulfur (Hemalatha et *al.*, 2013; Lavanya et *al.*, 2019). Fertilizers used in agricultural production systems include mineral fertilizers (urea, ammonium nitrate, sulfates and phosphates) and organic fertilizers (animal manure, biosolids and composts). Mineral fertilizers are mainly nitrate, ammonium, phosphate and potassium salts (Aoun et *al.*, 2010).

Fertilizers are widely used in agriculture as a quick and inexpensive source of plant nutrients, resulting in a considerable increase in crop productivity (Ahmad et *al.*, 2017). They often double or even triple yields (Sanginga & Woomer, 2009), while playing a key role in improving crop productivity and soil fertility, and their use has expanded considerably worldwide since the Green Revolution (FAO, 2017; Mylona et *al.*, 2018). The use of mineral fertilizers has long been an essential tool for compensating for nutrient losses and thus increasing yields (Amanullah et *al.*,2016; Wang et *al.*, 2020; Liu et *al.* ,2021) and have seen a significant increase in their use worldwide (Amanullah et *al.*,2016; Wang et *al.*, 2020). It is estimated that 30-50% of agricultural yields in different parts of the world are attributed to synthetic commercial fertilizers (Gupta et *al.*, 2014; Yousaf et *al.*,2017). However, the massive use inorganic fertilizers agriculture worldwide to ensure global food security has led to numerous health problems and irreversible environmental pollution (Kumar & Prakash, 2019). This use hardens the soil, reduces its fertility, pollutes air, water and soil, and depletes important soil nutrients and minerals, thus endangering the environment (Pahalvi et *al.*, 2022). These fertilizers are inherently persistent in nature and can deteriorate soil quality, the soil microbial communities that maintain the various nutrient cycles and the usefulness of groundwater for various purposes (Hossain et *al.* 2022).

As living beings depend on soil and water for their survival, they are also exposed to the health risks associated with these agrochemicals (Hossain et *al.* 2022). In recent decades, the use of fertilizers and pesticides has increased the exposure of farmers, agricultural workers and the general population to these chemicals (Sharma & Singhvi, 2017). Nitrates, phosphates and potassium salts, the main components of mineral fertilizers, are non-lipophilic substances and do not penetrate significantly percutaneously in the occupational environment. Orally, these elements are rapidly and significantly absorbed in duodenum and proximal small intestine. In the case of phosphates, 70-90% of the ingested dose is absorbed (Testud, 2011).

Accidental intoxications with inorganic fertilizers resulting in death and various pathological conditions have been reported in a variety of animals, including cattle (Horner 1982; Caldow & Wain 1991; Yeruham et *al.* 1997; Campagnolo et al. 2002), sheep and goats (East 1993; Campagnolo et *al.* 2002), and birds (Ley 1986; Raidal & Jaensch, 2006). Cases of acute fertilizer poisoning have also been reported in humans, resulting in acute renal failure, hyperkalemia and methemoglobinemia (Sato et *al.* 1999; Haung & Lin, 2004). In repeated-dose toxicity studies in which low percentages of inorganic fertilizers were administered to albino rats drinking water, anemia, leukopenia, liver and kidney damage and adverse reproductive effects observed (Ihedioha & Idika 2007 a and b).

Several epidemiological studies have reported that occupational exposure to higher levels of nitrate in employees can lead to increased body burdens of nitrate and nitrite in saliva (Moller, 1995). In addition, the risk of chronic exposure of workers to fertilizers in the occupational environment may have effects on fertility (Mallem et *al.*, 2015). The aim of our study is to conduct an experimental investigation into the toxic effects of a nitrogenous chemical fertilizer used in agriculture, and in particular to learn about the effects of increasing doses of Calcium Nitrate administered by gavage on certain parameters related to liver and kidney functions, and to compare alterations in certain haematological, biochemical and histological parameters in rats after fertilizer administration.

1. FERTILIZERS MINERALS

1.1 General information on fertilizers

Fertile agricultural land is essential for maintaining optimum productivity conditions to meet the food needs of the world's growing population. Soil health management is necessary to ensure sustainable agricultural production and the maintenance of biodiversity. Modern agriculture relies mainly on various inputs, such as pesticides and chemical fertilizers. The latter are essential for creating the right conditions for intensive, efficient farming; they have been used since ancient times, when nitrogen from animal and human excrement, phosphate from bones and potassium from ashes were added to the soil on an experimental basis. For thousands of years, the Egyptians used the abundant silt carried by the Nile to enrich and protect the soil. With the advent of the coal and petrochemical industries in the 19th , synthetic fertilizers appeared, followed by the three main elements (nitrogen, phosphorus, potassium). Fertilizers are substances that can penetrate all or part of a plant and promote its growth. They are combinations of minerals designed to provide plants with the additional nutrients they need for growth (Testaud, 2015). They are most often incorporated into the soil, but can also be transported by irrigation water (Darriet & Corbel, 2008).

There are two main types of fertilizer used in agriculture: organic fertilizers and mineral fertilizers. A third type fertilizer is organomineral. These are made up of mineral matter and a minimum of 25% organic substances of animal or plant origin. They are a mixture of mineral fertilizers and organic fertilizers that complement each other.

Organic fertilizers are essentially derived from plant matter (e.g. compost, manure), animal matter or human excrement, they are high-carbon compounds that enhance plant productivity and growth performance (Assefa & Tadess 2019). Organic fertilizers nevertheless often contain organic substances that can increase plant growth and yield by improving soil properties. They can increase the soil's capacity to retain water, thus plant access water (Yuniwati et al., 2023). They consist mainly of manure, crop residuesgreen manures, compost, etc. (Kéita, 1985; Diallo, 2002). Various organic materials have been recognized as soil conditioners and disease control agents, including control of brown spot disease and increase in bacterial by rice bran (Osunlaja, 1989) and the increase in plant growth and reduction in nematode population by oilcake (Khan & Saxena, 1997).

These organic materials undergo a series of transformations that decompose them, transform them into humus and then mineralize them. This series of transformations is carried out by micro-organisms.

Artificial fertilizers first appeared in the mid-19th . They are compounds containing the high concentration of nutrients required for plant growth. According to Bliefert and Perraud (1997), artificial or synthetic fertilizers in particular are those produced by factories or mines; in everyday language, term artificial fertilizer is used for most commercial mineral fertilizers. Fertilizers are man-made, marketed and used worldwide; they are derived from eruptive, sedimentary or saline rocks, or are obtained by synthesis or industrial processes. Fertilizers come in a variety of shapes and sizes, depending on the manufacturing process: granules, pellets, crystals, coarse or fine powders, etc. These chemical fertilizers provide plants with the nutrients they need: nitrogen, phosphorus and potassium. Secondary elements or trace elements contribute soil enrichment. They are also a source of secondary nutrients, including sulfur (Hemalatha et *al*., 2013; Lavanya et *al*., 2019). Fertilizers contain varying proportions of major (N, P, K, etc.) and minor (Zn, Mn, Fe, etc.) plant-essential elements, well as impurities and other non-essential elements.

1.2. Different types fertilizer chemical

In agriculture, the soil nutrients that improve productivity and are widely used are mineral fertilizers, the most important of which are nitrogen, phosphate and potassium fertilizers.

1.2.1. Fertilizers

Nitrogen (N) is one of the main constituents of plant and animal life, playing an active role in many metabolic processes and in the phenology of crops and vegetables (Khan et *al*., 2013), its deficiency in the soil limits plant growth. The main sources of N for fertilization urea, ammonium nitrate, ammonium sulfate, calcium nitrate and others.

(Goyal, 2012). In the soil, it exists in a mineral form (NO_3^- and NH_4^+), which can be assimilated by crops, and in an organic form. The latter, representing up to 90 % of total N in mineral soils (Olk, 2007), constitutes the soil's main nitrogen reserve. Mineralization of these compounds provides plants with an easily-assimilable source nitrogen.

1.2.1. 1. The nitrogen cycle

Figure 1 summarizes the nitrogen cycle in the biosphere. It is also characterized by the mobilization and immobilization of nitrogen compounds. Most reactions are catalyzed only by microfauna and microorganisms. Fixation occurs either through chemical oxidation, which renders the element assimilable (Ramade, 1998); or through the activity of specialized free-living or symbiotic microorganisms (Sharma & Ahlert, 1977; Luce et *al.*, 2011; Sullivan et *al.*,2014). The main N loss pathway in soil is denitrification, which is the reduction of NO3 - or NO2 - to N2O, NO and N2 (Drury et *al.*, 1992; Saggar et *al.*, 2013). It takes place under anaerobic conditions, with N2 and N2O being the dominant results under fully and partially anaerobic conditions respectively (Drury et *al.*, 1992). Denitrification is generally weak at pH below 4 (Postma et *al.*, 2006). It is triggered by denitrifying and anammox bacteria, which are anaerobic bacteria that derive their energy from the reduction nitrogen oxides (Strous et *al.*, 1999).

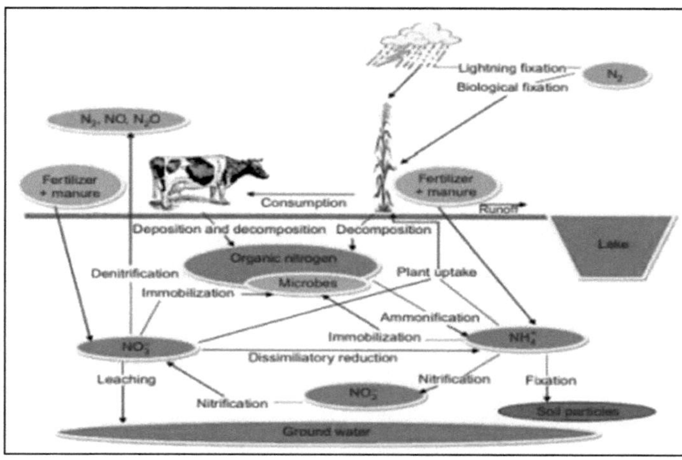

Figure 1: Nitrogen cycle in soil (Luce et *al.*, 2011)

1.2.1. 2. Use nitrogen fertilizers

The use of nitrogen fertilizers is one of the most important land management practices for improving crop and pasture productivity (Mazzetto et *al.*, 2020). After the Second World War, inorganic nitrogen fertilizers were introduced to improve crop productivity to meet the growing demands of the world's population in the 1950s. This led to the Green Revolution with the introduction of fertilizer varieties to achieve high yields in the 1960s. Over the past half-century, the doubling of crop yields is linked to a sevenfold increase in use of nitrogen fertilizers worldwide (Han et *al.*, 2016). Its use increased from 10.8 million metric tons in 1960 to 82 million metric tons in 2000, and is set to rise further to 249 million tons by 2050. The most widely used synthetic nitrogen fertilizers and their nitrogen contents are shown in **Table 1** (Cao et *al.*, 2017).

Table 1. Different types nitrogen fertilizers and their nitrogen contents (Cao et *al.*, 2017).

Type of fertilizer	Chemical formula	Nitrogen content (%)	Nitrogen form
Anhydrous ammonia	NH_3	82	NH_4^+
Ammonia	NH_3	22	NH_4^+
Ammonium	NH_4NO_3	34	50% NH_4^+ and 50% NO_3^-
Ammonium	$(NH_4)_2SO_4$	21	NH_4^+
Nitrogen solutions	$CO(NH_2)_2$, NH_4NO_3	30	75% NH_4^+ and 25% NO_3^-
Sodium nitrate	$NaNO_3$	16	NO_3^-
Urea	$CO(NH_2)_2$	46	NH_4^+
Calcium nitrate	$Ca(NO_3)_2$	17	NO_3^-
Diammonium phosphate	$(NH_4)_2HPO_4$	18	NH_4^+
Monoammonium phosphate	$NH_4H_2PO_4$	11	NH_4^+
Ammonium Phosphates	$(NH_4)_3PO_4$	15	NH_4^+

1.2.2. Phosphate fertilizers

Phosphorus is one of the three main nutrients essential for plant growth (Lazali et *al.*, 2020). It is a component of the enzymes needed for the energetic reactions involved in plant respiration and photosynthesis. It is also a component of certain lipids and numerous energy-binding compounds such ADP and ATP (Hopkins, 2003). It plays a crucial role in energy , plant metabolism and protein activation (Priya & Sahi, 2009). It is one of the main constituents of nucleic acids, joining nucleotides. It is also a constituent of plant membrane phospholipids (Sanchez et *al.*, 2009). Its deficiency in environment is a limiting factor for cultivation. This deficiency must be remedied by phosphate fertilization, which is necessary for plant growth and yield optimization (Talboys et *al.* ,2016). Globally, the main user of phosphorus is agriculture, accounting for around 90% of total global demand (Gupta et *al.*, 2014).

1.2.2.1. The phosphorus cycle

The P cycle is a dynamic transfer system between soil, plant and microbial biomass **(Fig. 2)** (Pierzynski et *al.*, 2000). Consequently, the cycle includes the processes of phosphorus uptake by the plant, transformation into organic forms and the chemical reactions of fixation and immobilization in the soil (Morel, 2002; Demers, 2008). Phosphorus in soil solution accounts for less than 1% of total phosphorus, but is the main source of phosphorus for plants (Pereda Campos, 2008). According to Morel (1996), it corresponds to phosphorus dissolved in the soil solution, in the ionized form of phosphoric acid H2PO4-, HPO4 2-, PO4 3-, the proportions of which vary according to the pH of the medium concerned. These forms are responsible for the phosphate nutrition of crops. According to Drouet (2010), microorganisms also play an important role in mineralization and immobilization processes, and an active phosphorus cycle can be discerned in the soil. This cycle is quite similar to that nitrogen. According to Eliard (1979), the phosphorus cycle is the only one of the main biogeochemical cycles to have no gaseous component.

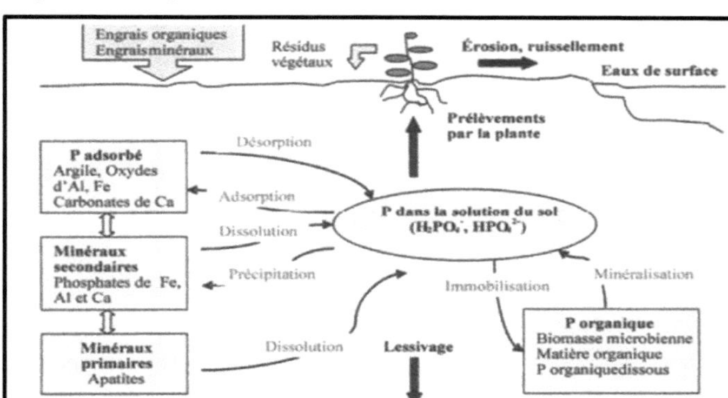

Figure 2: Phosphorus cycle (Pierzynski et *al.*, 2005)

1.2.2.2. The use of phosphate fertilizers

Phosphate mineral fertilizers enable high agricultural yields and help feed the world's population (Bindraban, 2020). They are widely used around the world (Hartley et *al.* ,2013; Taliman et *al.* ,2019) they provide one of the most important nutrients for agricultural production Indeed, around 150 million tonnes of phosphate rock are mined worldwide every year for the production PO4 fertilizers (Valadares et *al.* ,2017; Bhat et *al.* ,2019). Commercial phosphate fertilizers can contain two forms of phosphorus (orthophosphates), either an acid-reacting $H_2PO_4^{-1}$ form (triple superphosphate (TSP), monoammonium phosphate (MAP)) or a basic-reacting HPO_4^{-2} form (diammonium phosphate (DAP)) (Giroux, 1991). Phosphate fertilizers are not fully utilized by crops. Plants use only 5% to 30% of the added P (McLaughlin et *al.*, 1988), leaving 70% to 95% residual P. Over time, the soluble P from fertilizers accumulated in soils reacts with soil components, making it less and less available to plants (Stevenson, 1986).

1.2.3. Fertilizers potassium

K is an essential element in biological functioning in plants, used for the process of photosynthesis, carbohydrate and protein formation and migration, osmoregulation and enzyme regulation (Marschner, 1995; Hatterer & Kessler, 2014). Unlike nitrogen and phosphate, which are the basic building blocks proteins, nucleic acids, phospholipids, ATP, etc., potassium is not incorporated into plant molecules. Potassium exists predominantly as free element or as an absorbed limited cation, and can therefore be easily removed from the cell membrane of even the entire plant (Lindhauer, 1985). This high mobility in the plant explains potassium's main physiological characteristics: it is the main cation involved in charge neutralization and the most important active inorganic element (Clarkson & Hanson, 1980). Potassium plays an important role in increasing the rate of transport of nutrients (especially sucrose and amino acids) in the elaborated sap (Herlihy, 1989). Potassium deficiency affects several metabolic processes, such as the efficiency of photosynthesis, translocation and respiration, as well as various enzyme systems (Marschner, 1995; Mengel, 1997).

1.2.3.1. Potassium cycle

The potassium cycle is closely linked to the physico-chemical characteristics of the soil.

To reason potassium fertilization, it is necessary to know the dynamics of potassium in the soil. Potassium exists in the soil in four different forms: soluble, exchangeable, non-exchangeable (fixed) and structural K (Agbenin, 2020). Unlike nitrogen and phosphorus, potassium exists in both solid form (soil aggregates) and liquid form (soil solution) **(fig. 3).** Potassium in soil solution is fixed by humus or clay; potassium in minerals is released only very slowly. Under certain conditions, fixed potassium can be exchanged with the previous forms. Excessive K should be avoided, as it is subject leaching, especially in sandy soils. Similarly, there is competition for K in the soil between the plant and the clay-humus complex. To maintain an adequate level of exchangeable potassium the soil, it is advisable to apply more potassium to clay soils than to sandy soils, where it is preferable to divide up applications and synchronize them with plant needs. Agricultural exports and leaching (more significant in sandy soils with low organic matter and low cation exchange capacity) are the main causes of loss of this nutrient. The main entry routes are mineral fertilizers and organic inputs.

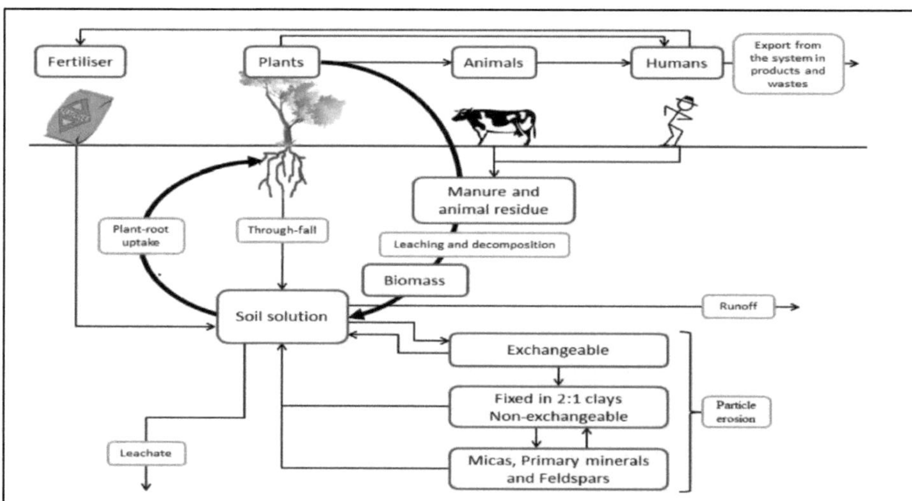

Figure 3. The potassium cycle in soils (adapted from Brady and Weil (2008)).

Brady, N.C., Weil, R.R., (2008). The Nature and Properties of Soil, 14 ed. Prentice Hall, New Jersey, USA, 965 pp.

1.2.3.2. The use of potassium fertilizers

According to Smil (1999), K fertilizers, unlike nitrogen (N) and phosphorus (P), are applied at a much lower rate, and less than 50% of the K removed by crops is replenished. Since 1980, the use of K fertilizers has increased by around 25%, and global demand for K fertilizers is set to rise further, from 23.8 Mt K in 2011 to 27.1 Mt K in 2015 (FAO, 2011) due to the targeted increase in global agricultural production. On a global scale, cereal crops (wheat, rice, corn) account for the largest share (37%) of K fertilizer use, followed by fruit and vegetables (), oilseeds (), sugar and cotton (11 %) and other crops (14%) (FAO, 2011). For most annual crops, the general practice is to apply K fertilizers at or before . Glauconite sand contains around 4-8% K2O and has been used as a source of potassium fertilizer worldwide (Majumder et *al.* 1995; Yadav et *al.* 2000). Potassium chloride (KCl), a natural mineral extracted from deep deposits, is the main form of K fertilizer used today. Potassium sulphate and potassium nitrate, both by-products of KCl extraction, are also available, but are more expensive. Organic fertilizers (animal manure, slurry, biogas residues or food waste) are also an important source K fertilizers in agriculture. The potash requirements of fertilizers can be determined by plant and soil analysis.

1.4. Consumption of fertilizers

Globally, the application of crop fertilizers has increased astronomically over the past 60 years (Gupta et al., 2014). Global consumption of the three main nutrient fertilizers, nitrogen, phosphorus and potassium reached 186.67 million tonnes in 2016, with an increase of 2.4% recorded in 2015 (FAO, 2017). Total demand for fertilizers (N, P, K) is rising steadily, as shown **in Figure 3**, increasing annually by 1.9 % on average from 2015 and is expected to reach 201.66 million tonnes in 2020 (FAO 2017). In Algeria, over the past two decades, the agricultural sector has been the driving force behind the country's economic growth, accounting for almost 13% of the working population (Bessaoud & al., 2019). Pesticides and fertilizers are now used in greater quantities compared to the past in order increase crop yields (Băcanu et *al.*, 2019).

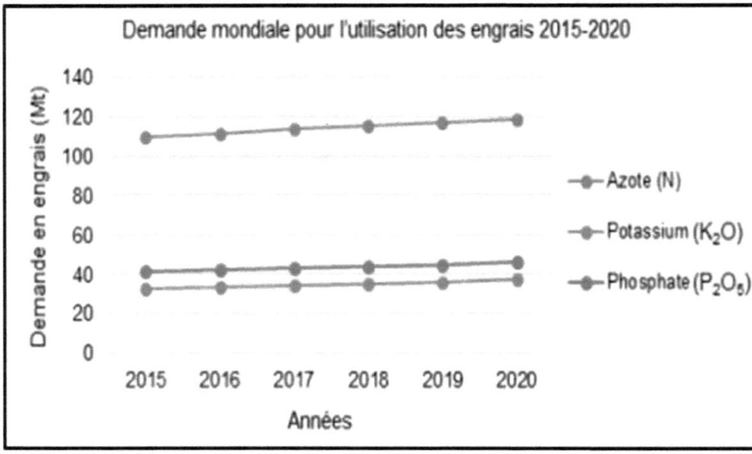

Figure 4: Global demand for the main fertilizers (N, P, K) from 2015 to 2020 (FAO 2017)

I.5. Impact of fertilizers

The massive use of fertilizers, while guaranteeing greater food security for mankind, has exerted and is currently exerting disastrous effects on the environment, in particular greenhouse gases (GHGs) (Davis et *al*., 2016) and ammonia (NH3) (Steffen et *al*., 2015), as well as nitrate leaching (Wang et al., 2019). Several researchers have reported that only a third to half of minerals applied as fertilizers are actually absorbed by plants, and that the remaining half or two-thirds generally become sources of soil, water and air pollution, causing serious problems for the environment and human health (Sheahan et *al*., 2017; Elahi et *al*., 2019; Li et *al*., 2019).

I.5.1. Impact of fertilizers on the environment

➢ **Impact of fertilizers on the air**

Excessive fertilizer application contributes to global warming through the emission of nitrous oxide (N2O) and nitrogen oxides (NOx), and the volatilization of ammonia (NH3) (Vitousek et *al*., 2009). Nitrous oxide (N2O) is a atmospheric tracer gas with a global warming potential 265 times that of carbon dioxide (CO2) over a 100-year period (Pachauri & Meyer, 2014), and is the main ozone-depleting agent of the 21st century (Portmann et *al*., 2012). Cropland is globally responsible for 66% of total anthropogenic

N2O emissions worldwide, and these emissions are expected to double by 2050 (Davidson & Kanter, 2014). High inputs of nitrogen fertilizers increase the magnitude of N2O emissions from cropping systems and are responsible for an estimated 54% of the total increase in global N2O emissions over the past decade (Tian et *al.*, 2019).

- **Impact of fertilizers on the soil**

Long-term application of mineral fertilizers and manure significantly alters soil chemical properties (Bünemann et *al.*, 2018; Chowdhury et *al.*,2019), for example soil reaction, salinity, cation exchange capacity, organic carbon content and soil nutrient status. Fertilization, depending on the method, affects the content of mainly absorbable forms of macronutrients and micronutrients in the soil. It also affects the long-term abundance of microorganisms and the structure and dynamics of the soil microbial community, as microorganisms are a sensitive component of an agricultural ecosystem (Lupatini et *al.*, 2017; Wang et *al.*, 2019). Balanced long-term organic and mineral fertilization has a generally positive effect on bacterial, actinobacterial and fungal populations. It increases their metabolic activity, which accelerates the decomposition of organic matter and the circulation of nutrients, and also increases their availability to plants (Elbl, 2019). On the other hand, unbalanced nutrient fertilization can cause microbial imbalance in the soil. This reduces biodiversity, biomass and microbiological activity. In particular, the use of high doses of nitrogen fertilizers reduces the number of certain types of bacteria (Zhou et *al.*, 2015; Zhong et *al.*, 2015). Excess nitrogen leads to toxic compounds, notably ammonia and nitrosamine, accumulating in the soil, reducing the presence of many groups of microorganisms and soil acidification (Barabasz et *al.*, 2002; Geisseler & Scow, 2014). However, long-term use mineral fertilizers is associated with changes in soil pH (Geisseler & Scow, 2014), which has important implications for bacterial community compositions (Delgado-Baquerizo et *al.*, 2018). Low soil pH increases the presence of fungi, including toxin-producing species of the genera Aspergillus, Fusarium and Penicillium. The metabolites they release, mainly mycotoxins, are toxic to other soil organisms (Francioli et *al.*, 2016).

- **Impact of fertilizers water**

Leaching of mineral elements and groundwater contamination, mainly by nitrates and dissolved reactive phosphorus, are major concerns worldwide. Nitrogen is an extremely mobile element in the soil when in the form of nitrate, whereas phosphorus is a very weakly mobile element in the soil (Haynes, 1985). Nitrate is easily and rapidly leached through the soil compartment if its removal by the crop is poorly synchronized with its availability; a large proportion is then lost through infiltration into groundwater (Trenkel, 2021), unlike ammonium, which is leached in much smaller quantities, as it is the absorbable form favoured by a large number of plants, in addition to its positive charge, which enables it to be easily retained by the soil's clay-humus complex thanks to its cation exchange capacity (Gagnon, 2009). As a result of the over-application of nitrogen and phosphate fertilizers on farmland, a large fraction of these elements is lost with runoff water or through leaching and percolation to the table. This leads to the proliferation of algae and aquatic plants rivers and lakes, resulting in eutrophication **(Fig.5)**. This can cause serious damage to drinking water quality and treatment, animal and human health, and aquatic ecosystems (Beusen et *al.*, 2016; Huang et *al.* 2017).

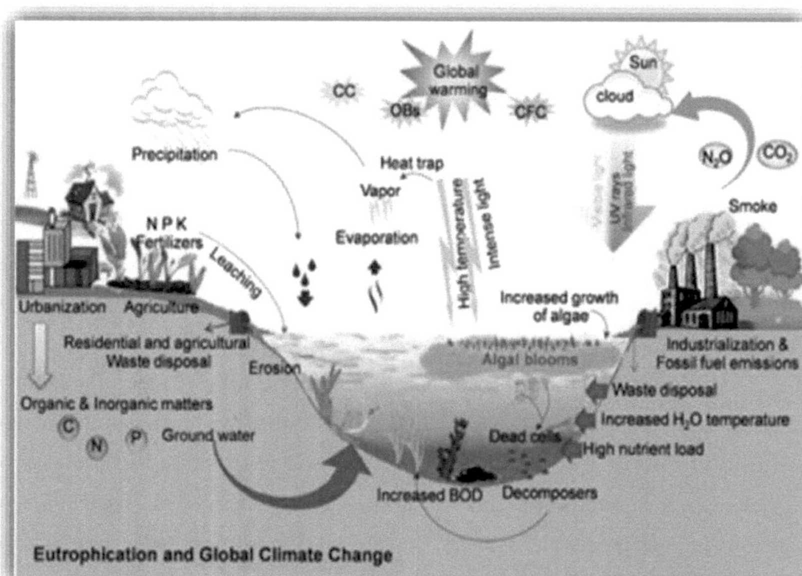

Figure 5: Negative effects nitrogen and phosphate fertilizer use on ecosystems. (Rastogi et *al.*, 2015)

I.5.2. Impact of fertilizers on human health

Fertilizers are one of the essential inputs for improving agricultural production. cumulative adverse effects of irrational fertilizer use not only intensify soil, water and atmospheric pollution, but have also posed a threat to biodiversity and human health. Chemical fertilizers can indirectly affect wildlife by disrupting food chains and habitats.

The main risk from inorganic chemical mineral fertilizers comes from the nitrogen components, which are present in most fertilizers. Nitrogenous components are nitrates NO3, ammonium NH4 (usually in the form ammonium nitrate NH4NO3) and urea. Even with low exposure, fertilizers can have serious consequences for the body, such as male infertility and cancer, but also serious consequences for the fetus. Drinking nitrate-contaminated water can be dangerous, especially for infants and pregnant women, as it can lead to a potentially serious disease called methemoglobinemia, or "blue baby syndrome" (Haung & Lin, 2004).

Professional fertilizer workers and farmers who handle and apply chemical fertilizers can be exposed to potentially hazardous chemical substances. This can occur through inhalation of fertilizer dusts or vapors, direct skin contact or accidental ingestion. Such exposure can lead health problems such as skin irritation, respiratory problems and neurological disorders. Some people may be sensitive to the chemicals in chemical fertilizers. Allergic reactions can occur when exposed to these chemicals, leading to symptoms such as skin rashes, itching, eye irritation, respiratory problems and respiratory allergies. In addition, the consumption of animals or plants poisoned by these same elements is a further danger to humans.

2. CALCIUM NITRATE

2.1. Description and physico-chemical properties of calcium nitrate

Calcium nitrate tetrahydrate (Leclaire & Monier, 1970) **(Tab. 2)** is one of the most widely used inorganic fertilizers, accounting for 4% all nitrogen fertilizers used . It is recommended for use on acid soils, as it acidifies soils less than many common nitrogen fertilizers (Pierre, 1928). It is a highly effective compound for plant nutrition, providing calcium and nitrogen. It appears most often as an ingredient in certain fertilizers or in granulated form. It has a high solubility (Boyd, 1981) and can also be found in solution versions (Rodriguez & Zea, 2015).

Table 2. Summary of the physico-chemical properties of calcium nitrate

Identification of product	Product name : CALNISOL
Use	Soluble fertilizer for crop fertigation.
Chemical formula	Ca (NO3)2 4H2O
Composition	• Total nitrogen (N): 15%. Nitric nitrogen (NO3): 15%. • Calcium oxide (CaO): .
Physical and chemical properties	Appearance (formulation): white crystals. - Odorless. - Water-soluble. - Solubility: 1212 g/l.
Detailed structure	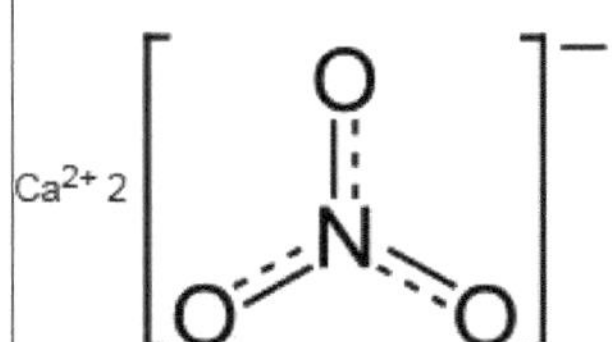
Toxicity	LD 50: 3900 mg/kg in rats
Molecular weight	236.15 g/mol.

2.2. Process for manufacturing calcium nitrate

Calcium nitrate is manufactured industrially by two different processes (Sauchelli, 1960). In the first process, ground limestone is reacted with nitric acid. After purification and the addition of ammonium , a hot solution produced by a spray-drying stage, the final product is a white, water-soluble crystal granulated at 15.5% nitrogen. The product will henceforth be known as calcium nitrate A. In another process, as a by-product in the manufacture of nitrophosphates. Rock phosphate is dissolved with nitric acid and calcium nitrate is separated from the mother liquid by cooling, crystallization and centrifugation (Killingmo & Luthman, 1968).

2.3. Use of calcium nitrate

Calcium nitrate is a salt with a variety commercial applications. It is used in concrete to prevent rebar corrosion (Justnes, 2002; Justnes & Nygaard, 1995) It is used in the production of fertilizers, explosives, matches, , incandescent mantles, and added to diesel as corrosion inhibitor (Ropp, 2013). Calcium nitrate is also used to prevent odors in wastewater collection and treatment (Weissenberger and *al.*, 2006).

2.4. The nitrate

Nitrate is a natural form of nitrogen and an integral part of the nitrogen cycle in the environment. Nitrate is formed from fertilizers, decomposing plants, manure and other organic residues. It is found in air, soil, water and food (particularly vegetables), and is produced naturally by the human body.

2.4. 1. nitrate in food

Nitrate is also used as a food additive, mainly as a preservative and antimicrobial agent. Sodium nitrate (E251) or potassium nitrate (E252) are commonly used in foods such as cheese and cheese products, raw and processed meats, edible casings, processed fish, fish , spirits and liqueurs. Plant species likely to accumulate nitrates include Brassica plants, green cereals (oats, wheat, rye and corn)sorghum and Sudan grass, corn, beets, sweet clover and raspberries (Wrathall, 2002). Vegetables and meat products are the main sources of nitrates in food, but small amounts may be present in fish and dairy products. Meat products may contain <2.7-945 mg nitrate/kg; dairy products generally contain <3-27 mg nitrate/kg. Food processing, fertilizer use and growing conditions (particularly soil temperature and daylight intensity) influence the amount of nitrates in vegetables. Crops such as beet, salad greens, radishes and spinach often contain nitrates at concentrations above 2,500 mg/kg, especially when grown in greenhouses (WHO, 2011).

2.4. 2. nitrate drinking water

In recent years, nitrate contamination of groundwater has been a growing problem worldwide. The highest nitrate concentrations in groundwater have been recorded in areas of intensive agricultural production (Filipović et al., 2013). By using nitrogen fertilizers, nitrates will be leached under the plant root zone and eventually reach groundwater (Oram, 2015), as the nitrate ion does not bind to the soil adsorption complex and is susceptible to flushing from the soil to groundwater (Sofilić, 2014). Nitrate-rich drinking water is associated with private wells, particularly wells (<15 m deep) in areas with permeable soils (Fewtrell, 2014).

Due to the increased use of synthetic nitrogen fertilizers and livestock effluents in intensive agriculture, vegetables and drinking water may contain higher concentrations of nitrates than in the past. The World Health Organization (WHO, 2006) specifies 50 mg/l nitrate in water as the maximum recommended concentration.

2.4. 3. Nitrate in the human body

There are two main sources of nitrate in the human body. One is the endogenous formation of nitric oxide (NO), which is spontaneously oxidized in blood and tissues to nitrite, and the other is diet. Wagner et al (1983) stated that diet is not the only source of nitrate, but that nitrate is synthesized endogenously. Consequently, the total amount of nitrate excreted in urine does not come from the diet, although during the period of high nitrogen intake, the contribution of endogenous nitrate biosynthesis to urinary nitrate is small. Research by Leaf et al (1989) endogenous biosynthesis from arginine.

2.5. Toxicity of nitrate

Nitrate exists widely in drinking water and food, indeed it inevitably enters the human food chain. Consequently, understanding the metabolism of ingested nitrate is a prerequisite to the proper assessment of the expected harmful and beneficial effects on human health. Ingested nitrate is primarily absorbed from the upper portion of the digestive tract. Absorption occurs rapidly, between one and three hours ingestion of food and beverages, and many researchers have found peak levels of nitrate in serum, saliva and urine (Bartholomew and Hill, 1984). Absorbed nitrate is rapidly distributed to the salivary glands and possibly to other exocrine glands, and may even pass into breast milk (Green et al, 1982). Nitrate reduction begins in the mouth with bacteria that produce the enzyme nitrate reductase, also found in mammalian gut microbes. Around 25% of ingested nitrate is converted to nitrogen dioxide (NO_2) in the mouth, and the acidic pH of the stomach promotes the formation nitrous acid, which can break down into various nitrogen oxides, including nitrogen dioxide (NO_2) and dinitrogen trioxide (N_2O_3), depending on the redox environment and gastric contents. The enterosalivary circulation of nitrate has been described after its absorption in the intestine. The salivary glands then recover around 20% of the circulating nitrate and secrete it into the oral cavity, where it is reduced to nitrite by the local microflora and restart the cycle (Fig.6) (Pereira et *al.*, 2013). In the stomach, ingested nitrite is converted into unstable nitrous acids, which spontaneously degrade into a variety of nitrogen compounds, mainly nitric oxide and N-nitroso compounds (Du et *al.*, 2007).

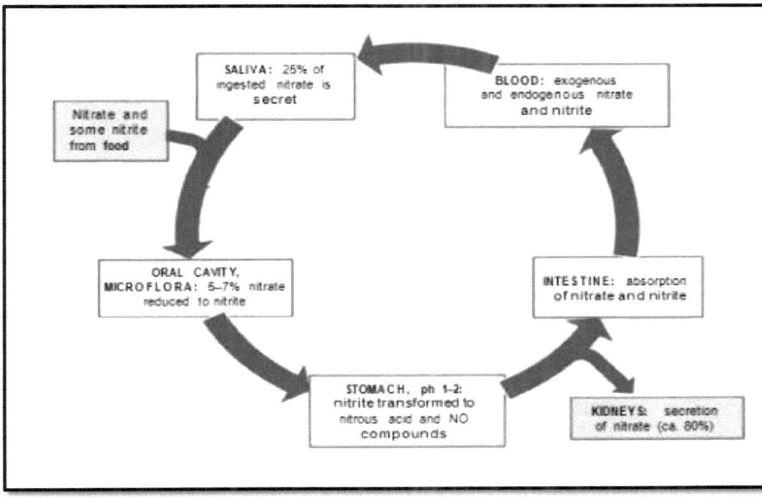

Figure 6. Nitrate metabolism (Hmelak Gorenjak & Cencič, 2013).

Nitrate is the precursor of nitric oxide (NO), which has a number of physiological functions, including neurotransmission (Garthwaite, 1991), immune regulation (Hibbs, 1991), vascular relaxation of smooth muscle (Moncada et *al.*, 1986) and inhibition of platelet aggregation (Radomski et *al.*, 1987). In addition, nitric oxide (NO) interacts with oxyhaemoglobin (HbO_2) to form methaemoglobin (metHb), and can react with secondary amines to produce N-nitro compounds that may be carcinogenic (Volkmer et *al.*, 2005). Moreover, the latter are hepatotoxic, causing fibrosis and tumors (Erkekoglu & Baydar, 2010).

.

3. MATERIALS AND METHODS

3.1. Hardware

3.1.1. Equipment chemical

In our study, we chose to define and test two of the most widely used fertilizers in Algeria, particularly in the Souk-Ahras region. The first step was a field survey of farmers and fertilizer sellers in the Souk-Ahras region (Taoura, Ouled Driss, Hnancha, Souk-Ahras). This approach led us to test two fertilizers that were on the list of fertilizers most used by farmers in the region. The first was a nitrogen fertilizer (calcium nitrate).

Calcium (Ca(NO 3) 2 -4H 2 O) marketed under the name Calnisol® (case no.: 10124- 37-5) purchased from a fertilizer company, Profert, Bejaia, Algeria. Calcium nitrate was dissolved in mineral water for oral administration; doses were administered according body weight.

3.1.2. Materials

This work was carried out on white male rats (Albino Wistar). These are nocturnal mammals from the order of rodents commonly used in research. Sexual maturity occurs between the fiftieth and sixtieth day after birth. A healthy rat can live from two and a half to three years, depending on environmental conditions.

3.1.2.1. Pets and maintenance conditions

Twenty-eight (28) male rats from the Institut Pasteur in Algiers, aged eight weeks and weighing 240±30 g. All were kept in our University's animal house and underwent an adaptation period of around three weeks to animal house conditions (temperature 23±2°c photoperiod 10/14). Rats were housed in plastic cages lined with wood shavings. The cages were cleaned and the bedding changed once every two days. The animals were fed a dry kibble diet of wheat, maize and barley from the food complex (Oued Fragha, Guelma, Algeria). Tap water was distributed to the animals in bottles, and renewed daily **(Fig. 7).**

Figure 7. Animal care packaging **(Original photo)**

3.1.2.2. Rat treatment

After the adaptation period (21 days), the rats were randomly divided into seven groups of seven rats each; the rats were treated for thirty (30) days as follows:

Group 1 (Control): the control group received tap water by gavage (*per os)*. **Group 2 (Calcium nitrate D1)**: rats were treated with calcium nitrate at a rate of **(200 mg/kg)** of body weight by oral gavage per day for 30 days.

Group 3 (Calcium nitrate D2): rats were treated with calcium nitrate **at (400 mg/kg**body weight by oral gavage daily for 30 days.

Group 4 (Calcium nitrate D3): rats were treated with calcium nitrate **at (800 mg/kg)** body weight by gavage for 30 days.

NB: Animals were weighed daily for the duration of treatment. Doses were chosen according to the literature.

3.1.2.3. Sacrifice and sample collection

➢ **Blood sampling**

At the end of the treatment period, the rats were left to fast overnight. They were sacrificed in the morning by decapitation, in the **LST2E laboratory**, and the blood was immediately collected in two different tubes:

One with EDTA: is immediately transported in a cooler to the medical analysis laboratory to determine the blood count formula (FNS).

The other with heparin: centrifuged at 3,000 rpm for 15 minutes, then the resulting serum is recovered and stored at -20°C for biochemical parameters.

- **Organ harvesting**

After dissection, the liver and kidneys were carefully removed, rinsed 0.9% NaCl and weighed. They were stored in two ways: one half was stored in the freezer for subsequent preparation of the homogenate used to assay oxidative stress parameters, the other half was fixed in 10% formalin solution for histological sections. **Figure 8** shows the various stages of the experimental protocols.

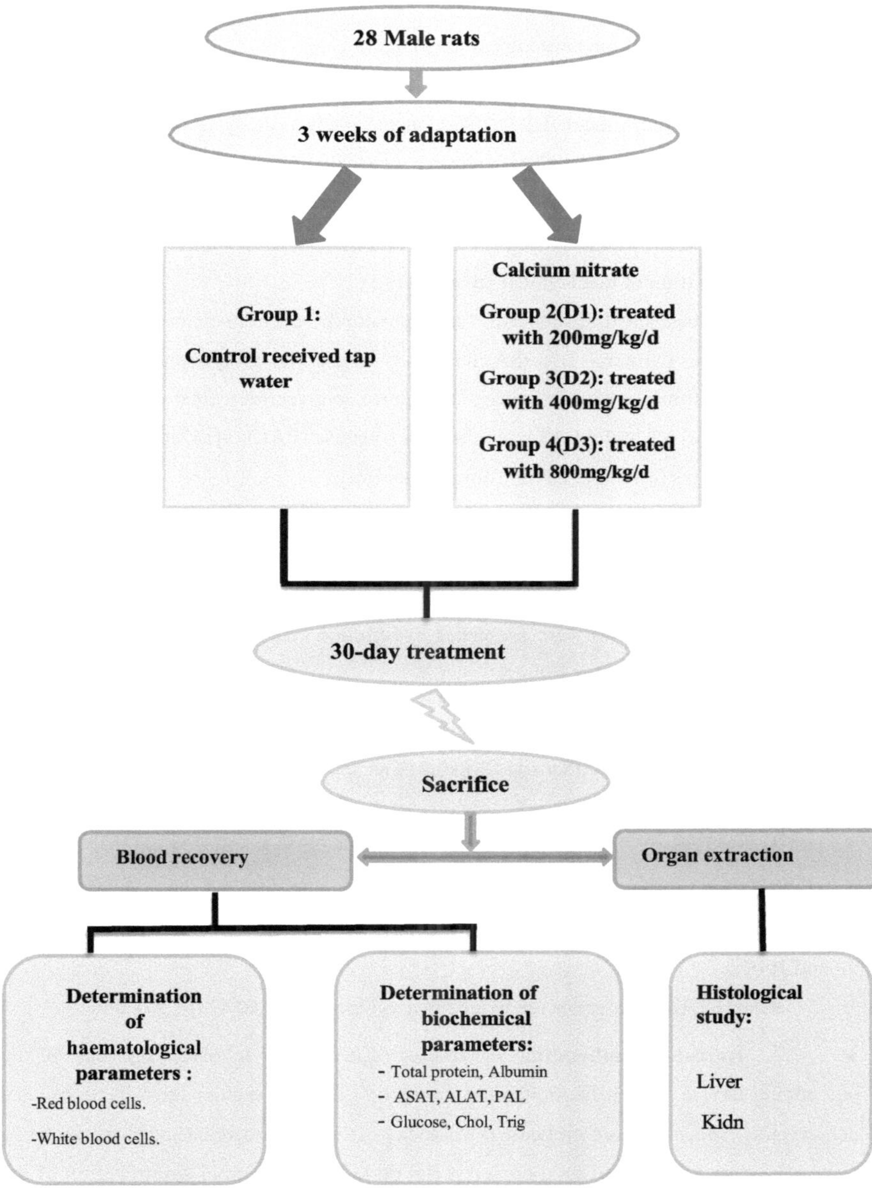

Figure 8. Summary diagram of experimental protocol

3.2. Methods

3.2.1. Determination of haematological parameters :

The blood count formula (FNS) was performed by an automatic hematology analyzer (Mindray Coulter, Inc, model: BS - 30C.), in the medical analysis laboratory of EPSP Taoura, to determine the following parameters: red blood cell (RBC), white blood cell (WBC), hemoglobin (Hb), hematocrit (HT), lymphocytes (LY), monocytes (MO), mean red cell volume (MCV) and mean corpuscular hemoglobin concentration (MCHC).

3.2.2. Determination of biochemical parameters :

The various biochemical parameters are measured spectrophotometrically using a spectrophotometer (Mindray BA-88A) and ready-to-use kits (Diagnopharm kits, Bouira, Algeria) to determine levels of glucose, cholesterol, triglycerides, total protein, albumin, transaminases (ASAT and ALAT), alkaline phosphatase (PAL), urea, creatinine and uric acid. Assays are performed at EPSP Taoura laboratory.

3.2.3. Histological study

Histological sections were taken at the private anatomical pathology laboratory (Dr. Amoura) in Souk-Ahras. Using the technique described by Houlot (1984). The harvested organs (livers and kidneys) were preserved in formalin (10%) and subjected to various stages of histological study as follows:

- **Fixation**: The aim of fixation is to structural integrity. It carried out by immersion, bathing the specimens in formalin. The specimens are then removed from the formalin and cut transversely. The specimens were then placed in special cassettes.
- **Dehydration**: This stage consists of removing the water. Samples are then dehydrated for a minimum of 12 hours, using an automatic device that automatically and progressively transfers samples to ethanol baths of increasing concentration (70%, 80%, 90 and 100%).
- **Clarification**: cassettes are placed in a xylene bath at 40°C for 2, 3 times.
- **Kerosene embedding of tissues**: This stage involves the use of an embedding device, which facilitates the production of thin, regular cross-sections . To achieve this, specimens are embedded in blocks of liquid kerosene (under the effect of heat). They are then placed on a cooled metal plate to ensure rapid solidification of the kerosene block containing the tissue. A microtome is used to slice the block to the desired thickness, producing sections 2 to 5 μm thick. The resulting sections are then

spread out on slides.

- **Dewaxing**: to remove the kerosene, the slides are placed on a hot plate (at 45-60°C) for 15 min, obtain liquefaction and thus kerosene removal.
- **Rehydration**: Slides are immersed in alcohol baths of decreasing strength from 100° to 50° alcohol), then distilled water.
- **Staining with combination of hematoxylin and eosin (HE) dyes**: The Hematoxylin & Eosin technique requires the presence of acid alcohol (100 ml 70% ethyl alcohol + 50 ml HCl acid), ammonia water (100 ml distilled water + 2 ml ammonia) and eosin solution.

Colouring is carried out according to the following steps:

- ✓ Dewax and moisten slides tap water, then rinse distilled water.
- ✓ Soak in Harris Hematoxylin bath (15 minutes), which stains basophilic structures (nuclei) purplish blue.
- ✓ Distinguish between the components of acid-alcohol cuts
- ✓ Then place the slides in a tap water bath and check differentiation under the microscope.
- ✓ Blue in ammonia water bath.
- ✓ Soak in an Eosin bath (15 seconds to 2 minutes) to stain acidophilic structures (cytoplasm) pink. All baths are separated by rinsing with tap water.
- ✓ The preparations were then dried and observed under an optical microscope (brand name OPTIKA B-290, Italy).

3.2.4. Study statistics

Results were represented as mean plus or minus standard deviation (MOY ± SD), with comparisons between groups made after analysis of variance (ANOVA).

Data analysis was performed using SPSS statistical software (Version 23.). Differences are considered as :

- *Significant when ($p \leq 0.05$).
- ** Highly significant when ($p \leq 0.01$).
- ***Very highly significant when ($p \leq 0.001$)

4. RESULTS

4.1. Effects of treatment on variations in some physiological parameters.

4.1.1. Body weight changes in rats.

Table 3 shows the variations in body weight of the different groups. Throughout the treatment period, monitoring of animal body weight indicates that animals treated with calcium nitrate at a dose of (800 mg /kg) showed a highly significant increase (P<0.01) compared with the control group. This body weight gain was accompanied by an increase in the percentage of weight gain compared with the control group **(Tab.3).**

Table 3. Body mass changes in control and calcium nitrate-treated rats at 200 mg/kg (G2), calcium nitrate at 400 mg/kg (G3), calcium nitrate at 800 mg/kg (G4), after 30 days of treatment.

Parameters	Experimental groups			
	Parameters	Calcium nitrate(G2)	Calcium nitrate(G3)	Calcium nitrate(G4)
Initial mass(g)	256,14±6,28	257,14±4,52	257,29±3,95	256,86±5,55
Final weight (g)	289,43±6,24	289,29±3,35	291,86±5,58	300,14±3,67**
Weight gain (%)	33,29±10,06	32,14±5,34	34,57±3,50	43,29±6,73*

Values are expressed as mean ± SD (n = 7rats); *; P< 0.05: Significant difference compared with control; **; P <0.01: Highly significant difference compared with control.

4.1.2. Organ weight variations: liver and kidney.

Absolute (PA) and relative (PR) target organ weights, in this case liver and kidney, were estimated in control and treated rats for 30 days.

The results show that administration of calcium nitrate at 800 mg/kg body weight significantly (p<0.01) increased absolute and relative weights. There was also a highly significant increase in absolute and relative liver mass (+17.50%; +13.33% respectively) in the group treated with the dose.Concerning absolute and relative kidney weight, analysis of the results obtained shows that calcium nitrate had no effect on absolute and relative kidney weight compared with the control batch **(Tab. 4).**

Table 4. Changes in absolute (g) and relative (g/100g CP) liver and kidney weights in control and calcium nitrate-treated rats at 200 mg/kg (G2), calcium nitrate at 400 mg/kg (G3), calcium nitrate at 800 mg/kg (G4), after 30 days of treatment.

Experimental groups Parameters Control (G1) Calcium nitrate(G7) calcium(G5)			Calcium nitrate	Calcium nitrate(G6)
Absolute liver weight	7,83±0,46	8,06±0,36	8,77±0,66**	9,20±0,48***
%	—	+2,94	+12,00	+17,50
Relative liver weight	2,70±0,14	2,79±0,14	3,01±0,24*	3,06±0,17***
%	—	+3,33	+11,48	+13 ,33
Absolute kidney weight	0,67±0.05	0,68±0,04	0,69±0,04	0,72±0,05
%	—	+1,49	+2,98	+7,46
Relative weight of kidneys	0,23±0,01	0,24±0,02	0,24±0,02	0,24±0,02
%	—	+4,35	+4,35	+4,35

Values are expressed as mean ± SD (n = 7rats); * ; P< 0.05 : Significant difference compared to control; ** ; P <0.01 : Highly significant difference compared to control; *** ; P< 0.001 Very highly significant difference compared to control.

4.2. Effects of treatment on haematological parameters

The results of the haemogram or blood count (FNS) are **shown in Table 5**.

Our results reveal that treatment with calcium nitrate for 30 days causes a very highly significant (P< 0.001) decrease in hemoglobin, hematocrit and red blood cells (-6.59%; -6.22%; -12.92% respectively) in the G4 group. On the other hand, a highly significant increase (P< 0.01) in VGM in the same group

Our results also show that treatment of rats with both fertilizers for 30 days produces a highly significant (P < 0.01) increase in white blood cells in the higher dose group. However, an increase in lymphocytes was observed in the G4 group (+19.82%) compared with the control batch.

In addition, statistical analyses show that there were no significant variations in CCMH and PLT in all treated groups compared controls.

Table 5. Changes in haematological parameters in control rats and rats treated with calcium nitrate 200 mg/kg (G2), calcium nitrate 400 mg/kg (G3), calcium nitrate 800 mg/kg (G4), after 30 days of treatment.

Parameters	Experimental groups			
	Control (G1)	Calcium nitrate (G2)	Calcium nitrate (G3)	Calcium nitrate (G4)
GB ($x10^3/\mu l$)	7,54±0,83	7,47±0,56	8,47±0,54*	8,89±0,38**
%	—	-0,93	+12,33	+17,90
GR ($x10^6/\mu l$)	8,90±0,41	8,67±0,51	8,27±0,37**	7,75±0,26***
%	—	-2,58	-7,08	-12,92
HB (g/dl)	16,84±0,53	16,56±0,53	15,90± 0,75*	15,73±0,34***
%	—	-1,66	-5,58	-6,59
Hte (%)	45,36±0,67	44,93±1,07	43,89±0,94**	42,54±0,49**
%	—	-0,95	-3,24	-6,22
PLT ($x10^3/\mu l$)	578,43±44,55	572,86±81,83	559,29±47,42	556,14±43,23
%	—	-0,96	-3,31	-3,85
VGM (fl)	50 ,99±2,56	51,98±3,10	53,15±1,70	54,97±2,36**
%	—	+1,94	+4,24	+7,81
CCMH (pg)	37,14±1,25	36,86±1,19	36,25±2,05	36,98±0,89
%	—	-0,75	-2,40	-0,43
LY (%)	3,38±0,57	3,36±0,54	3,63±0,67	4,05±0,21**
%	—	-0,59	+7,40	+19,82
MO (%)	1,11±0,10	1,10±0,09	1,06±0,11	1,02±0,10
%	—	-0,90	-4,50	-8,11

Values are expressed as mean ± SD (n = 7rats); * ; $P< 0.05$: Significant difference compared to control; ** ; $P <0.01$: Highly significant difference compared to control; *** ; $P< 0.001$ Very highly significant difference compared to control.

4.3. Effects of treatment on biochemical parameters

Oral gavage of rats with calcium nitrate led to disturbances in biochemical metabolism. The results observed after plasma determination of biochemical parameters are shown in **tables 6.7.8.**

4.3.1. Effects of treatment on biochemical markers of liver function

The effect of oral gavage of calcium nitrate for 30 days on enzyme activity reflecting liver function is shown in **Table 6.**

- **Total protein**: Results indicate that administration of calcium nitrate at 400 and 800 mg/kg body weight to rats resulted in a highly significant ($p<0.001$) decrease in plasma total protein content (-14.47%; -16.31%). Rats treated only with the low dose showed no significant difference compared with control rats **(Fig. 9).**
- **Albumin: Figure 9** shows that treatment of rats with calcium nitrate (800 mg/kg) caused a highly significant decrease in albumin (-5.61%) and a highly significant decrease in albumin (-6.92%) compared to the control group.
- **Transaminases**: Our results reveal that exposure of rats to calcium nitrate at 800mg/kg caused a very highly significant ($p< 0.001$) increase in ASAT (+17.47%), accompanied by a highly significant ($p< 0.01$) increase in ALAT (+16.46%) compared with controls, and no significant difference in rats treated with the 200mg/kg dose compared with control rats **(Fig. 10).**
- **Alkaline phosphatase**: Treatment of rats with calcium nitrate at 800mg/kg increased alkaline phosphatase enzyme activity (+17.05%) to a highly significant level ($p<0.01$) **(Fig. 11).**
- **Total bilirubin**: The results shown in Fig. 11 show a significant increase ($p<0.05$) in total bilirubin levels in G4 rats compared with control rats **(Fig. 11).**

➢

Table 6. Changes in serum biomarkers of liver function in control and calcium nitrate-treated rats at 200 mg/kg (G2), calcium nitrate at 400 mg/kg (G3), calcium at 800 mg/kg (G4), after 30 days of treatment.

Parameters	Experimental groups			
	Control (G1)	Calcium nitrate (G5)	Calcium nitrate (G6)	Calcium nitrate (G7)
Total protein (g/l)	71,17±5,18	64,37±8,17	60,87±4,09***	59,56±3,12***
%	—	-9,55	-14,47	-16,31
Albumin (g/l)	36,00±1,58	35,04±1,14	33,98±1,70*	33,51±1,27**
%	—	-2,67	-5,61	-6,92
AST (U/L)	114,64±16,29	111,73±10,30	123,37±5,35	134,67±5,29**
%	—	-2,54	+7,61	+17.47
ALAT (U/L)	67,60±9,03	63,40±9,50	73,41±4,45	78,73±5,69**
%	—	-6,21	+8,59	+16,46
PAL (U/L)	117,16±5,46	118,31±10,02	129,67±5,41	137,14±7,40**
%	—	+0,98	+10,68	+17,05
Bilirubin T (mg/l)	1,20±0,10	1,18±0,13	1,26±0,14	1,37±0,16*
%	—	-1,58	+4,90	+13,80

Values are expressed as mean ± SD (n = 7rats); *; P< 0.05: Significant difference compared to control; **; P <0.01: Highly significant difference compared to control; ***; P< 0.001 Very highly significant difference compared to control.

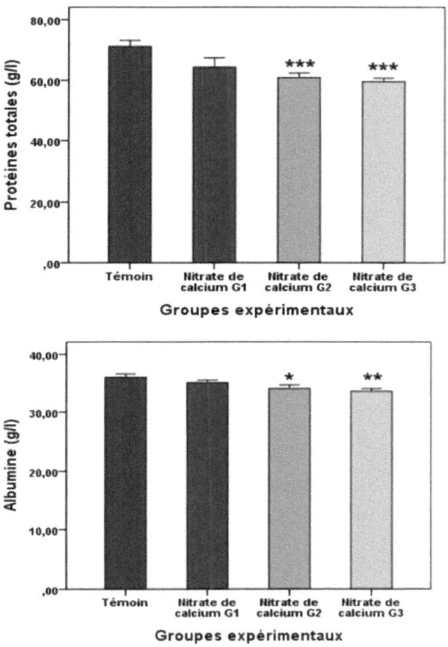

Figure 9. Changes in plasma total protein and albumin concentrations in control and calcium nitrate-treated rats at 200 mg/kg (G2), calcium nitrate at 400 mg/kg (G3), calcium at 800 mg/kg (G4), after 30 days of treatment.

*** ; p< 0.001 Very highly significant difference compared with control; * ; p< 0.05 : Significant difference compared with control; ** ; p <0.01 : Highly significant difference compared with control.

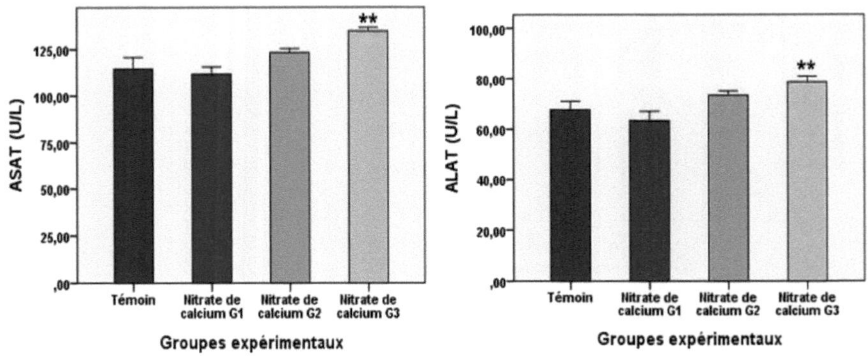

Figure 10. Changes in AST and ALAT enzyme activities in control and calcium nitrate-treated rats at 200 mg/kg (G2), calcium nitrate at 400 mg/kg (G3), calcium nitrate at 800 mg/kg (G4), after 30 days of treatment.

* ; $p< 0.05$: Significant difference compared with control; ** ; $p<0.01$: Highly significant difference compared with control.

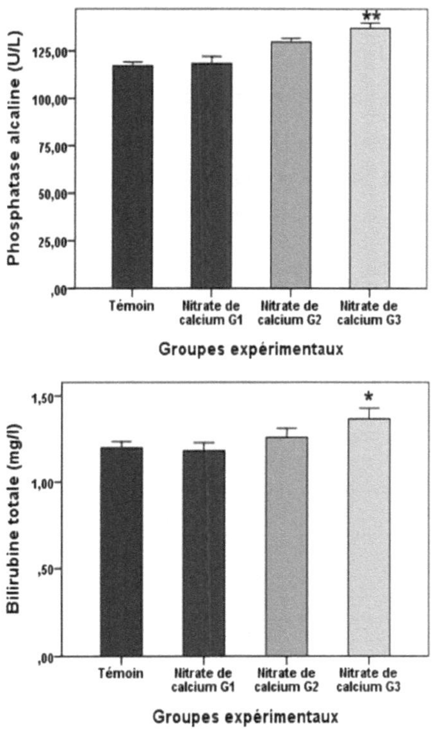

Figure 11. Changes in PAL enzyme activity and plasma total bilirubin concentration in control and calcium nitrate-treated rats at 200 mg/kg (G2), calcium nitrate at 400 mg/kg (G3), calcium nitrate at 800 mg/kg (G4), after 30 days of treatment.

* ; p<0.05 : Highly significant difference compared with control; *** ; p< 0.001 Very highly significant difference compared with control.

4.3.2. Effects of treatment on carbohydrate metabolism and lipid metabolism

The results shown in Table 8 illustrate the plasma levels associated with energy metabolism (glucose, cholesterol and triglycerides).

➢ **Glucose:** According to our results **(Fig. 12),** there was a highly significant variation ($p < 0.01$) in the calcium nitrate-treated group (G4) (+11.61%). The comparison between the groups treated with the low dose (200mg/kg) shows no significant difference from the control group.

➢ **Cholesterol:** The results reveal a highly significant increase in plasma cholesterol levels in the 800 mg/kg calcium nitrate group. Statistical analysis revealed no significant difference between the 200 mg/kg fertilizer groups **(Fig. 13).**

➢ **Triglycerides:** Plasma triglyceride concentration was highly significantly (P< 0.001) elevated by calcium nitrate G4 (+26.47%) compared with the control group. This concentration increased significantly (p< 0.05) in animals treated with calcium nitrate 400 mg/kg (+17.65) compared with untreated rats. There was no variation in the G2 group **(Fig. 13).**

Table 7. Changes in carbohydrate and lipid metabolism parameters in control and calcium nitrate-treated rats at 200 mg/kg (G2), calcium nitrate at 400 mg/kg (G3), calcium nitrate at 800 mg/kg (G4), after 30 days of treatment.

Parameters	Experimental groups			
	Control(G1)	Calcium nitrate(G2)	Calcium nitrate(G3)	Calcium nitrate(G4)
Blood glucose (g/l)	1,12± 0,07	1,16±0,09	1,21±0,05*	1,25±0,05**
%	—	+03,57	+08,04	+11,61
Cholesterol (g/l)	0,96±0,08	0,96±0,07	1,04±0,05*	1,11±0,04***
%	—	+00 ,00	+08,34	+15,62
Triglycerides (g/l)	0,68±0,13	0,67±0,07	0,80±0,03*	0,86±0,04**
%	—	-01,47	+17,65	+26,40

Values are expressed as mean ± SD (n = 7rats); *; P< 0.05: Significant difference compared to control; **; P <0.01: Highly significant difference compared to control; ***; P< 0.001 Very highly significant difference compared to control.

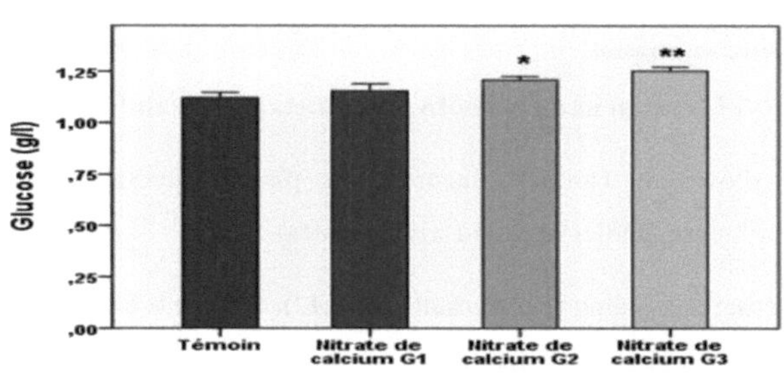

Figure 12. Changes in plasma glucose concentrations in control and calcium nitrate-treated rats at 200 mg/kg (G2), calcium nitrate at 400 mg/kg (G3), calcium at 800 mg/kg (G4), after 30 days of treatment.

* ; p < 0,05 : Significant difference compared to control ; ** ; p <0,01 : Highly significant difference compared to control ; *** ; p< 0,001 Very highly significant difference compared to control

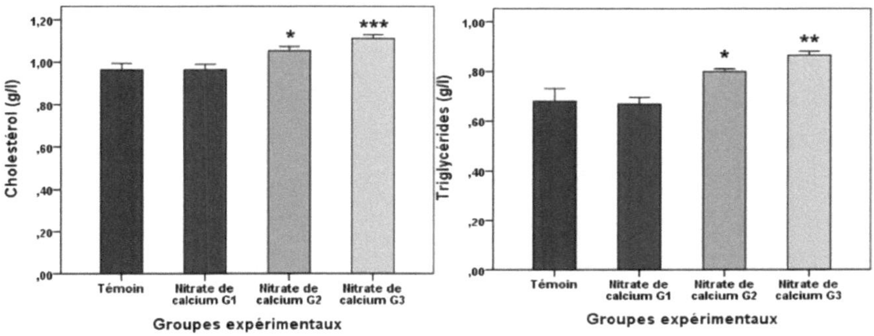

Figure 13. Changes in plasma cholesterol and triglyceride concentrations in control and calcium nitrate-treated rats at 200 mg/kg (G2), calcium nitrate at 400 mg/kg (G3), calcium nitrate at 800 mg/kg (G4), after 30 days of treatment.

* ; p < 0.05 : Significant difference compared with control; ** ; p <0.01 : Highly significant difference compared with control; *** ; p< 0.001 Very highly significant difference compared with control.

4.3.3. Effect of treatment on biochemical markers of renal function

The results obtained after assaying plasma markers associated with renal function (urea, creatinine and uric acid) at the end of experiment are presented in **Table 8**.

➢ **Urea:** Our results show that serum urea content increased highly significantly (P≤0.01) in rats treated with calcium nitrate (+07.50%). No significant difference was recorded in animals intoxicated with fertilizers at 200 and 400 mg/kg/day compared with controls **(Fig. 14)**.

➢ **Creatinine: According** to the results in **figure 14**, there was a highly significant (P<0.01) increase in plasma creatinine concentration (+18.48%; +11.21%) in rats treated with calcium nitrate (G4). There was no significant difference in creatinine levels in rats treated with both fertilizers (G2) and (G3) compared with the control group (G1).

➢ **Uric acid:** No significant variation was recorded in calcium nitrate-treated rats compared with the control batch **(Fig. 15)**.

Table 8. Changes in serum biomarkers of renal function in control and treated rats after 30 days of treatment

Parameters	Experimental groups			
	Control(G1)	Calcium nitrate(G5)	Calcium nitrate(G6)	Calcium nitrate(G7)
Urea (mg/dl)	40,20±02,72	40,50±02,66	41,58±02,26	43,49±02,41**
%	—	+00,00	+05,00	+07,50
Creatinine (mg/dl)	1,17±1,00	1,19±0,09	1,22±0,08	1,30±0,06**
%	—	+01,71	+04,27	+11,11
Uric acid (mg/dl)	4,75±0,54	4,59±0,49	4,94±0,44	5,31±0,44
%	—	-03,41	+03,94	+11,94

Values are expressed as mean ± SD (n = 7rats); P <0.01: Highly significant difference compared with control.

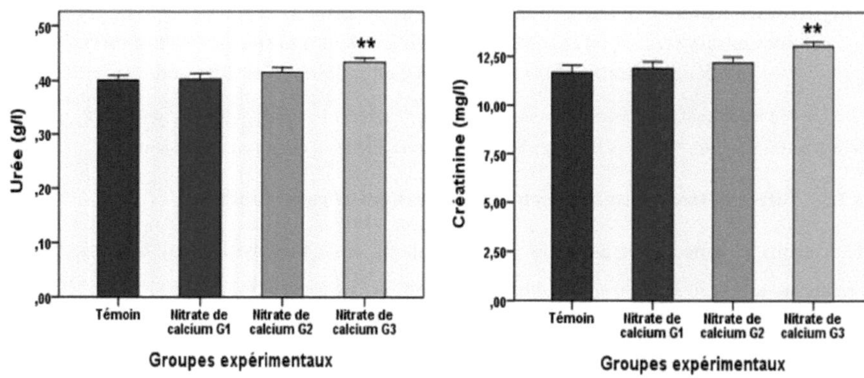

Figure 14. Changes in plasma urea and creatinine concentrations in control and calcium nitrate-treated rats at 200 mg/kg (G2), calcium nitrate at 400 mg/kg (G3), calcium nitrate at 800 mg/kg (G4), after 30 days of treatment.

** ; p <0.01 : Highly significant difference compared with control.

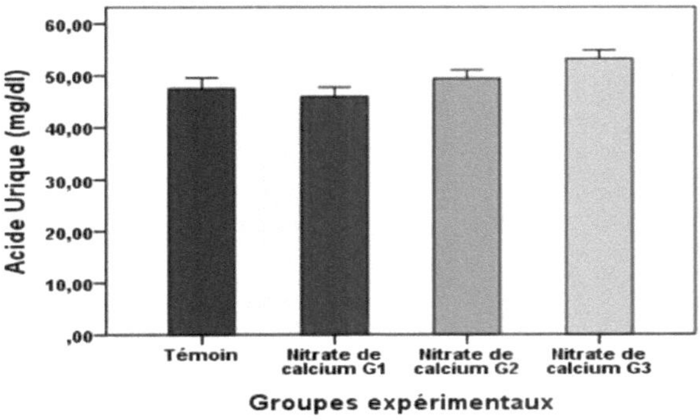

Figure 15. Changes in plasma uric acid concentrations in control and calcium nitrate-treated rats at 200 mg/kg (G2), calcium nitrate at 400 mg/kg (G3), calcium nitrate at 800 mg/kg (G4), after 30 days of treatment.

4.4. Histological study

4.4.1. Histological study of liver

Figure 16 shows the histopathological changes in the livers of male rats after exposure calcium nitrate at the indicated doses (200, 400 and 800 mg/kg) for 30 days, compared with control rats. Microscopic examination of liver tissue (x100) from untreated rats (Fig.15. A. B), indicates that the liver parenchyma has a normal arrangement of hepatocytes with visible capillary sinusoids. No histological damage was observed in the livers of the calcium nitrate-treated group at the low dose (200 mg/kg) compared with the control group. Liver microscopy (x100) of rats exposed to calcium at 400 and 800 mg/kg revealed considerable dilatation and vascular congestion, hepatocyte degeneration, hemorrhage and lipid vacuolization. There was also a mononuclear infiltration compared with the control group **(Fig. 15. C. D. E. F).**

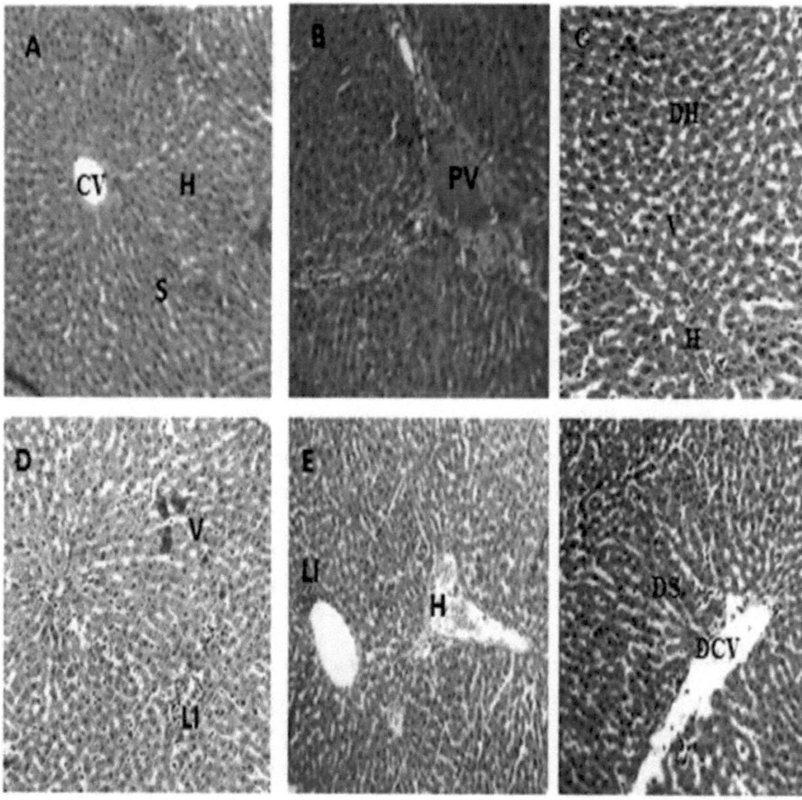

Figure 16. Histological sections of H & E-stained liver tissue from control rats and rats treated with calcium nitrate for 30 days. In control rats, the sections normal hepatic structure **(A, B)** with hepatocytes **(H)**, centrilobular vein **(CV)** and sinusoids **(S)**. In rats treated with calcium nitrate, the liver tissue severe alterations **(C.D.E.F)**.

DH, hepatocyte degeneration; **V**, vacuolization; **H**, hemorrhage; **LI**, leukocyte infiltration; **CVD**, centrilobular vein dilatation; **DS**, sinusoid dilatation;

4.4.2. Histological study of kidneys

Microscopic observation of kidney tissue (Gr. x 100) from control rats showed normal renal parenchyma with well-defined glomeruli and renal tubules **(Fig. 17.A).** In contrast, animals treated with calcium nitrate at the high doses (400 and 800 mg/kg) showed histopathological damage, including glomerular atrophy associated with dilatation of Bowman's space, and infiltration of inflammatory cells. In contrastintra-glomerular hemorrhage and vascular congestion were clearly observed in the kidney treated with calcium nitrate **(Fig. 17. B. C. D).**

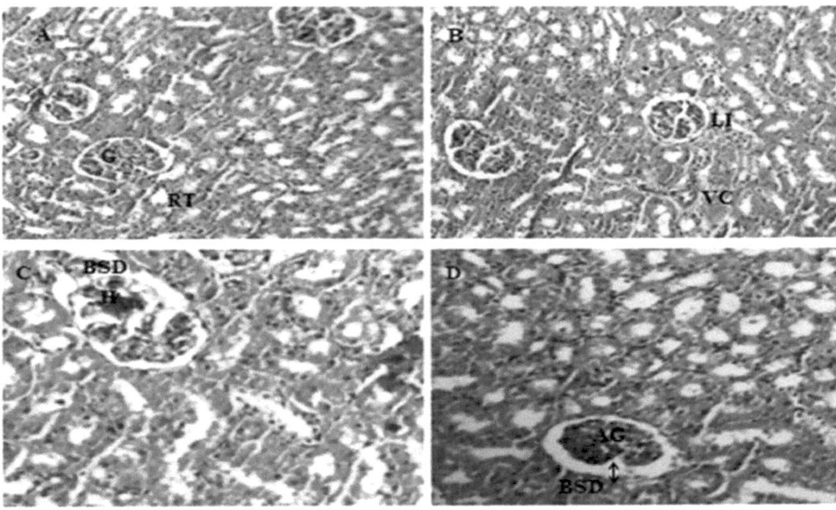

Figure 17. Histological sections of H & E-stained kidney from control rats and rats treated with calcium nitrate for 30 days. In control rats, the sections show normal kidney tissue with glomerular (G) and distal and proximal tubular (A) structure. In rats treated with calcium nitrate, microscopy shows alterations in renal tissue (B.C.D) (Gr. x 100).

AG, Glomerular atrophy; DSB Dilatation of Bowman's space; CV, Vascular congestion IL, Lymphocytic infiltration; H, Hemorrhage.

5. DISCUSSION

The extensive use of synthetic fertilizers has become a health problem affecting all mammals. The main causes of nitrate contamination of the environment are organic and chemical fertilizers (WHO, 2006). The chemical components of fertilizers have influenced the digestion and absorption process, as well as the metabolic utilization of ingested food (Weil & McCollister, 1963; OECD, 2000). Nitrates can leach into groundwater from soil-applied fertilizers or animal excrement, contaminating drinking water supplies (Lockhart, 2013). Nitrate toxicity poses a serious threat to animals and humans. In this context, we assessed the potential toxicity of a nitrogen fertilizer, calcium nitrate, in rats.

The study of body weight is an important criterion for assessing the toxic effects of xenobiotics, as variations in body weight can be attributed to altered biochemical reactions induced by toxicity (Mansour & Mossa, 2010). Under our experimental conditions, consumption of calcium nitrate-contaminated water causes a physiological imbalance marked essentially by a significant increase in body weight. This was explained by the accumulation of pollutant in adipose tissue, despite nitrate known to have an effect on body weight causing an increase in protein catabolism which may be necessary to meet energy requirements during detoxification (Al Ayed, 2000; Nujić & Habuda-Stanić, 2017). These data at odds with the work of Rouag et *al.* (2020), who found a decrease in body weight after exposure of rats to sodium nitrate. On the other hand, the work carried out by Messaadia et *al.* (2013) shows no significant change in the physiological growth of ammonium nitrate-treated rats, compared with control group. Similarly, Gueroui and Kechrid (2016) showed that the administration of silver in drinking water at 20 mg/l for 3 months in male rats did not disrupt their growth rate.

Organ weight estimation is a good indicator in toxicological research for assessing the adverse effects of xenobiotics (Bailey et *al.,* 2004, Sellers et *al.*,2007). Our results showed that 30 days' exposure to calcium nitrate caused a considerable increase in relative liver weight, implying that calcium nitrate consumption can cause hepatomegaly. This hypertrophy may explained by the overexcitation of detoxification mechanisms in response to the harmful effect of nitrate. Furthermore, our results are in agreement with those of (Lee et *al.*, 2020) who reported an increase in the hepatosomatic ratio males receiving 1000 mg/kg/day of ammonium nitrate. Similarly (Bouaziz-Ketata et *al.* 2014) reported an increase in relative liver weight In rats treated with NaNO3 for 7 weeks.

Among the biochemical changes produced by calcium nitrate treatment in rats was a considerable drop in total blood protein levels. Adult rats treated with sodium nitrate in drinking water at concentrations of 550 mg/L for four months recorded similar results (Azza et *al.*, 2011). In addition, the drop in serum total protein is mainly due to the effects of nitrate on the liver, either through necrotic alterations or other mechanisms (Anthony et *al.*, 1994). Administration of calcium nitrate to rats for 30 days resulted in elevation of transaminases (AST, ALT), PAL and bilirubin, as well as lowering of total proteins, particularly albumin. As these markers are often considered crucial in determining liver damage, this suggests liver cell damage. Our results are in line with information previously published by (Messaadia et *al.*, 2013; Fouad et *al.* ,2017; Kattiaa et *al.*, 2017). Furthermore, an increase in ALT and ASAT levels in the blood evoked liver damage and impaired function. These enzymes are the most sensitive indicators directly involved in the extent of cellular damage and toxicity, as previously discussed. They are diffused into the blood after cellular injury because they are cytoplasmic in nature (krim et *al.* ,2013). Increased activity of these enzymes in the serum of rats receiving calcium nitrate may be associated with necrosis and altered membrane permeability leading leakage from liver tissue. Previous experiments have found that rats given ammonium nitrate (Messaadia et *al.* ,2013), or sodium nitrate (Delgadoa et *al.* ,2018) had higher levels of ASAT and ALAT.

Our results clearly demonstrated a noticeable rise in glucose concentration in calcium nitrate-treated rats, in line with the findings of (Azzez et *al.*, 2011; Delgadoa et *al.* ,2018). The increase in blood glucose levels is probably explained by the stimulation of the glycogenolysis phenomenon, which allows glucose to be released by glycogen phosphorylase under action of amylase (Hijmans et *al.*, 2014). Numerous European studies have attempted establish a strong link between nitrate concentration in drinking water and insulin-dependent diabetes by increasing blood glucose while decreasing hepatic glycogen (IDD). In comparison, data are scarce and contradictory (Moltchanova et *al.*

,2004). The results obtained reveal a highly significant increase in serum concentrations of cholesterol and triglycerides, and this hyperlipidemia can be explained by hypothyroidism (Jublanc & Bruckert, 2004; Pearce, 2004). In animals, experimental studies have shown that inorganic nitrate is a short-term goitrogenic agent, causing hypertrophy of the epithelial cells of the thyroid gland (Gatseva & Argirova, 2008). These results correlate with the work of (Boukerche et *al*., 2007; Messaadia et *al,* 2013).

Creatinine, urea and uric acid are waste products produced by protein metabolism and eliminated by the kidneys (Tootian et *al*., 2012), generally used as indicators of renal function. The present results also revealed an increase in urea, creatinine and uric acid in response to calcium nitrate toxicity. Similar results were obtained by Messaadia et *al.* (2013). Increased serum urea, uric acid and creatinine levels suggested altered renal function. These alterations can be attributed to changes in reabsorption threshold, renal blood flow and glomerular filtration.

The results of the present work indicated that calcium nitrate induces histopathological alterations in the liver. The liver may show congestion of veins, mononuclear cell infiltration, cytoplasmic vacuolization of hepatocytes and fatty degeneration. Similarly, previous reports have demonstrated changes in adult rat liver tissue (Bouaziz-Ketata et *al.* 2014; Kattaia et al. (2017); Delgadoa et *al.*
,2018). The current results showed that treatment with au resulted in renal tubule dilatation, glomerular atrophy and leukocyte infiltration. Anwar & Mohamed (2015) reported glomerulus atrophy and renal blood vessel congestion in kidneys of rats treated 500 mg/L NaNO3 for 4 and 6 weeks.

6. REVIEW AND OUTLOOK

The extensive use synthetic fertilizers in agriculture has major consequences for the environment and human health. The consequences of these fertilizers on soil quality, biodiversity, water quality and ecosystem health are cause for concern. In addition, consumers face health risks due to the presence of chemical residues in fruit and vegetables grown with these fertilizers. Fruits and vegetables may have had their nutritional quality reduced due to the excessive use of chemical fertilizers, meaning that consumers may not derive as much nutritional value from their consumption. In addition, consumers may be exposed to toxic substances present in chemical fertilizers, which can lead to health problems such hormonal imbalances, developmental disorders and carcinogenic effects.

The results of this study indicate that high-dose oral gavage of fertilizers is capable of inducing adverse effects on visceral organs such as the liver and kidneys in rats, resulting in increased biochemical and enzymatic markers of liver and kidney function, which are further revealed by histopathological evidence.

The administration of calcium nitrate to rats caused disturbances in general growth of the rats, as evidenced by a significant increase in body weight accompanied by an increase in absolute and relative liver and kidney weights.

Our results obtained after treating rats with calcium nitrate at increasing doses of 200, 400 and 800 mg/kg can be summarized as follows:

- Increased body weight with increased relative and absolute liver weights
- Alterations in hematological parameters are reflected in a decrease in the number red blood cells, hemoglobin and hematocrit, and an increase in mean corpuscular volume, indicating the onset of macrocytic anemia;
- Disturbances in biochemical parameters and hepatic, renal and metabolic biomarkers confirm the hepatotoxic and nephrotoxic effect of calcium nitrate;
- Alterations in liver and kidney tissue are clearly revealed in histological sections.

Our results confirm the toxicity of fertilizers at different doses in rats on liver and kidney function.

Looking ahead, it would be interesting to develop this research work to deepen our knowledge. It is hoped to :

✓ Evaluate the toxicity of fertilizers over longer periods with other doses of fertilizers (calcium) highlight chronic harmful effects.

✓ Study detoxification capacities by measuring parameters such as nitric oxide radical, SOD, GR, ratio (GSH/GSSG, etc.) and stress proteins: Heat Shock Protein (HSP).

✓ Investigate the potential for detoxification of fertilizers by certain medicinal plants and test their phytotherapeutic properties on a laboratory model;(Rats or other).

✓ Further research into the fertilizer's mode of action in vivo with more in-depth studies (genotoxicity studies to look for DNA damage, or molecular biology studies to find out whether DNA damage is triggered directly).

REFERENCES

Agbenin, J. O. (2020). The Environmental Chemistry of Soils and Sediments. Principles and Applications: 277-313.

Ahmed, M., Rauf, M., Mukhtar, Z., & Saeed, N. A. (2017). Excessive use of nitrogenous fertilizers: an unawareness causing serious threats to environment and human health. Environmental Science and Pollution Research, 24(35), 26983-26987. https://doi.org/10.1007/s11356-017-0589-7

Al-Ayed, M. I. (2000). Toxicity of drinking water with different nitrate levels. Journal-Egyptian German Society of Zoology, 31(A), 197-210.

Amanullah, A.; Iqbal, A.; Ali, S.; Fahad, S.; Parmar, B. (2016). Nitrogen source and rate management improve maize productivity of smallholders under semiarid climates. *Front. Plant Sci. 7*, 1773. https://doi.org/10.3389/fpls.2016.01773.

Anthony, M. L., Gartland, K. P., Beddell, C. R., Lindon, J. C., & Nicholson, J. K. (1994). Studies of the biochemical toxicology of uranyl nitrate in the rat. Archives of toxicology, 68(1). https://doi.org/10.1007/s002040050028

Anwar, M. M., & Mohamed, N. E. (2015). Amelioration of liver and kidney functions disorders induced by sodium nitrate in rats using wheat germ oil. Journal of Radiation Research and Applied Sciences, 8(1), 77-83. https://doi.org/10.1016/j.jrras.2014.11.004

Aoun, M., El Samrani, A. G., Lartiges, B. S., Kazpard, V., & Saad, Z. (2010). Releases of phosphate fertilizer industry in the surrounding environment: Investigation on heavy metals and polonium-210 in soil. Journal of Environmental Sciences, 22(9), 1387-1397. https://doi.org/10.1016/S1001-0742(09)60247-3

Assefa, S., & Tadesse, S. (2019). The principal role of organic fertilizer on soil properties and agricultural productivity-a review. Agri Res and Tech: Open Access J, 22(2), 556192. https://doi.org/10.19080/ARTOAJ.2019.22.556192

Azzez OH, Mahmood MB & Hassan. (2011). Effect of nitrate poisoning on some biochemical parameters in rats. Iraqui J Vet Sci, 25 (2): 47-50.

Băcanu, C.S., Cristina, S.D., Iuliana Manuela, D.I. & Silvius, S. (2019). Agricultural production, soil quality and fertilizer used in Braila County, Romania. Research in

Agriculture and Agronomy. https://doi.org/10.5171/2019.970358.

Bailey, S. A., Zidell, R. H., & Perry, R. W. (2004). Relationships between organ weight and body/brain weight in the rat: what is the best analytical endpoint. Toxicologic pathology, 32(4), 448-466. https://doi.org/10.1080/01926230490465874

Barabasz, W., Albinska, D., Jaskowska, M., & Lipiec, J. (2002). Biological effects of mineral nitrogen fertilization on soil microorganisms. Polish Journal of Environmental Studies, 11(3), 193- 198.

Bartholomew, B., & Hill, M. J. (1984). The pharmacology of dietary nitrate and the origin of urinary nitrate. Food and Chemical Toxicology, 22(10), 789-795. https://doi.org/10.1016/0278- 6915(84)90116-9

Bessaoud, O., Pellissier, J. P., Rolland, J. P., & Khechimi, W. (2019). Rapport de synthèse sur l'agriculture en Algérie (Doctoral dissertation, CIHEAM-IAMM). p. 82.

Beusen, A. H., Bouwman, A. F., Van Beek, L. P., Mogollón, J. M., & Middelburg, J. J. (2016). Global riverine N and P transport to ocean increased during the 20th century despite increased retention along the aquatic continuum. Biogeosciences, 13(8), 2441-2451. https://doi.org/10.5194/bg-13-2441-2016

Bhat, J. A., Shivaraj, S. M., Singh, P., Navadagi, D. B., Tripathi, D. K., Dash, P. K., ... & Deshmukh, R. (2019). Role of silicon in mitigation of heavy metal stresses in crop plants. Plants, 8(3), 71. https://doi.org/10.3390/plants8030071

Bindraban, P. S., Dimkpa, C. O., & Pandey, R. (2020). Exploring phosphorus fertilizers and fertilization strategies for improved human and environmental health. Biology and Fertility of Soils, 56(3), 299-317. https://doi.org/10.1007/s00374-019-01430-2

Bliefert C., and Perraud R., (1997). Environmental chemistry: Air, Water, Soil, Waste. 1ère éd. Spain. 477p

Bouaziz-Ketata, H., Salah, G. B., Salah, H. B., Marrekchi, R., Jamoussi, K., Boudawara, T., ... & Zeghal, N. (2014). Nitrate-induced biochemical and histopathological changes in the liver of rats: ameliorative effect of Hyparrhenia hirta. Biomedical and Environmental Sciences, 27(9), 695- 706. https://doi.org/10.3967/bes2014.105

Boukerche, S., Aouacheri, W., & Saka, S. (2007). The toxic effects of nitrates: a biological study in humans and animals. In Annales de biologie Clinique (Vol. 65, No. 4, pp. 385-391). https://doi.org/10.1684/abc.2007.0136

Boyd, C. E. (1981). Solubility of granular inorganic fertilizers for fish ponds. Transactions of the American Fisheries Society, 110(3), 451-454. https://doi.org/10.1577/1548- 8659(1981)110%3C451:SOGIFF%3E2.0.CO;2

Bünemann, E. K., Bongiorno, G., Bai, Z., Creamer, R. E., De Deyn, G., de Goede, R., ... & Brussaard, L. (2018). Soil quality-A critical review. Soil Biology and Biochemistry, 120, 105-125. https://doi.org/10.1016/j.soilbio.2018.01.030

Caldow, G. L., & Wain, E. B. (1991). Urea poisoning in suckler cows. The Veterinary Record, 128(21), 489-491.

Campagnolo, E. R., Kasten, S., & Banerjee, M. (2002). Accidental ammonia exposure to county fair show livestock due to contaminated drinking water. Veterinary and human toxicology, 44(5), 282-288.

Cao, Y., Halane, M. K., Gassmann, W., & Stacey, G. (2017). The role of plant innate immunity in the legume-rhizobium symbiosis. Annu. Rev. Plant Biol, 68, 535-561.

Carvalho, F. P. (2006). Agriculture, pesticides, food security and food safety. Environmental science & policy, 9(7-8), 685-692. https://doi.org/10.1016/j.envsci.2006.08.002

Chowdhury, S.P., Babin, D., Sandmann, M., Jacquiod, S., Sommermann, L., Sørensen, S. J., ... & Grosch, R. (2019). Effect of long-term organic and mineral fertilization strategies on rhizosphere microbiota assemblage and performance of lettuce. Environmental microbiology, 21(7), 2426-2439. https://doi.org/10.1111/1462-2920.14631

Clarkson, D. T., & Hanson, J. B. (1980). The mineral nutrition of higher plants. Annual Review of Plant Physiology, 31: 239-298.

Conijn, J. G., Bindraban, P. S., Schröder, J. J., & Jongschaap, R. E. E. (2018). Can our global food system meet food demand within planetary boundaries? Agriculture, Ecosystems & Environment, 251, 244-256. https://doi.org/10.1016/j.agee.2017.06.001

Darriet, F., & Corbel, V. (2008). Influence of NPK fertilizers oviposition of Aedes aegypti. *Parasite*, *15*(1): 89-92. https://doi.org/10.1051/parasite/2008151089

Davidson, E. A., & Kanter, D. (2014). Inventories and scenarios of nitrous oxide emissions. Environmental Research Letters, 9(10), 105012 http://dx.doi.org/10.1088/1748-9326/9/10/105012

Davis, K. F., Gephart, J. A., Emery, K. A., Leach, A. M., Galloway, J. N., & D'Odorico, P. (2016). Meeting future food demand with current agricultural resources. Global Environmental Change, 39, 125-132. https://doi.org/10.1016/j.gloenvcha.2016.05.004

Delgado, M. G., Zamora, A. G., Gonsebatt, M. E., Mata, E. M., Vargas, G. G., Rincón, E. C., & Morales, R. P. (2018). Subacute intoxication with sodium nitrate induces hematological and biochemical alterations and liver injury in male Wistar rats. Ecotoxicology and Environmental Safety, 166, 48-55. https://doi.org/10.1016/j.ecoenv.2018.09.060

Delgado-Baquerizo, M., Oliverio, A. M., Brewer, T. E., Benavent-González, A., Eldridge, D. J., Bardgett, R. D., ... & Fierer, N. (2018). A global atlas of the dominant bacteria found in soil. Science, 359(6373), 320-325. https://doi.org/10.1126/science.aap9516

Demers, I. (2008). Formes et disponibilité du phosphore de composts utilisés comme amendements de sols agricoles, for the degree of Master of Science (M. Se.), Laval (Québec), 92p.

Diallo, L. (2002). Effect of urea and manure on maize yields. Mémoire d'Ingénieur du Développement Rural/Option Agronomie. IDRfUPB. Burkina .54p.

Drouet, TH. (2010). Pédologie, BING-F-302. Ed. Lagev, 140p.during incubation on the desorption of phosphorus from soil. Soil Sci.350-355.

Drury, C. F., Findlay, W. I., & McKenney, D. J. (1992). Nitric oxide and nitrous oxide production from soil: water and oxygen effects. Soil Science Society of America Journal, 56(3), 766-770. https://doi.org/10.2136/sssaj1992.03615995005600030015x

Du, S. T., Zhang, Y. S., & Lin, X. Y. (2007). Accumulation of nitrate in vegetables and its possible implications to human health. Agricultural Sciences in China, 6(10), 1246-1255. https://doi.org/10.1016/S1671-2927(07)60169-2

East, N. E. (1993). Accidental superphosphate fertilizer poisoning in pregnant ewes. Journal of the American Veterinary Medical Association, 203(8), 1176-1177.

EFMA (1997). Fertilizers - An Introduction. European Fertilizers Manufacturers Association (EFMA). http://www.efma.org/introduction/index.asp . (accessed on Dec. 10, 2005).

Elahi, E., Weijun, C., Zhang, H., & Nazeer, M. (2019). Agricultural intensification and damages to human health in relation to agrochemicals: Application of artificial intelligence. Land use policy, 83, 461-474. https://doi.org/10.1016/j.landusepol.2019.02.023

Elbl, J., Makovâ, J., Javorekovâ, S., Medo, J., Kintl, A., Lošák, T., & Lukas, V. (2019). Response of microbial activities in soil to various organic and mineral

J.B Baillière, Paris. 344p

Erkekoglu, P., & Baydar, T. (2010). Evaluation of the protective effect of ascorbic acid on nitrite-amendments as an indicator of soil quality. Agronomy, 9(9), 485 https://doi.org/10.3390/agronomy9090485.

ELIARD, J.L. (1979). Manuel d'agriculture générale, base de la production végétale. Ed and nitrosamine-induced cytotoxicity and genotoxicity in human hepatoma line. Toxicology mechanisms and methods, 20(2), 45-52. https://doi.org/10.3109/15376510903583711

FAO (2017). World fertilizer trends and outlook to 2020, Food and Agriculture Organization of the United Nations. Rome.

FAO (2017). *The Future of Food and Agriculture-Trends and Challenges*; FAO: Rome, Italy, Available online: **http://www.fao.org/3/i6583e/i6583e.pdf** (accessed on 1 December 2020).

Fewtrell, L. (2004). Drinking-water nitrate, methemoglobinemia, and global burden of : a discussion. *Environmental health perspectives*, *112*(14), 1371-1374.

Filipović, V., Petošić, D., & Nakić, Z. (2013). Presence of nitrates in groundwater:

Sources and processes Prisutnost nitrata u podzemnim vodama: Izvori i procesi. Hrvatske vode, 21(84), 119- 128.

Fouad, SS., Mohi-Eldin, M., Haridy, M. & Khalil, A. (2017). Ameliorative effects of ascorbic acid (Vit. C) against sodium nitrite toxicity in albino rats: hematological, biochemical and histopathological studies. American-Eurasian J Toxicol Sci, 9 (1): 01 06. https://doi.org/10.5829/idosi.aejts.2017.01.06.

Francioli, D., Schulz, E., Lentendu, G., Wubet, T., Buscot, F., & Reitz, T. (2016). Mineral vs. organic amendments: microbial community structure, activity and abundance of agriculturally relevant microbes are driven by long-term fertilization strategies. Frontiers in microbiology, 7, 1446. https://doi.org/10.3389/fmicb.2016.01446

Gagnon, J. (2009). Impact of different forms nitrogen (Urea, NH4+, NO3-) on plant growth and fertilizer leaching. Proceedings of the Session de Formation sur la Nutrition Minérale des Plants Forestiers, DRF-MRNF, Québec, QC, Canada, 72.

Garthwaite, J. (1991). Glutamate, nitric oxide and cell-cell signalling in the nervous system.

factor for thyroid dysfunction in children and pregnant women living in rural Bulgarian areas. International Journal of Hygiene and Environmental Health, 211(5-6), 555-559 https://doi.org/10.1016/j.ijheh.2007.10.002

Geisseler, D., & Trends in neurosciences, 14(2), 60-67. https://doi.org/10.1016/0166-2236(91)90022-M

Gatseva, P. D., & Argirova, M. D. (2008). High-nitrate levels in drinking water may be a risk **Scow, K. M. (2014).** Long-term effects of mineral fertilizers on soil microorganisms-A review. Soil Biology and Biochemistry, 75, 54-63. https://doi.org/10.1016/j.soilbio.2014.03.023

GIROUX, M. (1991). Comparaison des sources et des modes d'apport du phosphore et d'azote sur le rendement et la nutrition minérale du maïs, Agrosol IV, (1) : 43 - 53.

Gouel, C., & Guimbard, H. (2019). Nutrition transition and the structure of global food demand. American Journal of Agricultural Economics, 101(2), 383-403. https://doi.org/10.1093/ajae/aay03

Goyal, M. R. (2012). Management of drip/trickle or micro irrigation. CRC Press

Green, L. C., Wagner, D. A., Glogowski, J., Skipper, P. L., Wishnok, J. S., & Tannenbaum, S.
R. (1982). Analysis of nitrate, nitrite, and [15N] nitrate in biological fluids. Analytical biochemistry, 126(1), 131-138. https://doi.org/10.1016/0003-2697(82)90118-X

Gueroui, M., & Kechrid, Z. (2016). Evaluation of some biochemical parameters and brain oxidative stress in experimental rats chronically exposed to silver nitrate and the protective role of vitamin E and selenium. Toxicological research, 32(4), 301-309. http://dx.doi.org/10.5487/TR.2016.32.4.301

Gupta, D. K., Chatterjee, S., Datta, S., Veer, V., & Walther, C. (2014). Role of phosphate fertilizers in heavy metal uptake and detoxification of toxic metals. Chemosphere, 108, 134- 144. https://doi.org/10.1016/j.chemosphere.2014.01.030

Gupta, D. K., Chatterjee, S., Datta, S., Veer, V., & Walther, C. (2014). Role of phosphate fertilizers in heavy metal uptake and detoxification of toxic metals. Chemosphere, 108, 134-144. https://doi.org/10.1016/j.chemosphere.2014.01.030

Han, M., Wong, J., Su, T., Beatty, P. H., & Good, A. G. (2016). Identification of nitrogen use efficiency genes in barley: searching for QTLs controlling complex physiological traits. Frontiers in Plant Science, 7, 1587. https://doi.org/10.3389/fpls.2016.01587

Hartley, T. N., Macdonald, A. J., McGrath, S. P., & Zhao, F. J. (2013). Historical arsenic contamination of soil due to long-term phosphate fertiliser applications. Environmental Pollution, 180, 259-264. https://doi.org/10.1016/j.envpol.2013.05.034

Hatterer, A. and Kessler, H., "POTASSIUM", Encyclopædia Universalis [online], accessed June 22, 2014. http://www.universalis.fr/encyclopedie/potassium

Haynes, R. J. (1985). Principles of fertilizer use for trickle irrigated crops. Fertilizer research, 6(3), 235-255. https://doi.org/10.1007/BF01048798

Hemalatha, S., Radhika, K., Maragatham, S., & Praveena Katharine, S. (2013). Influence of long-term fertilization on soil fertility: A review. Research and Review: Journal of Agriculture and Allied Sciences, 2(3), 30-36.

Hibbs Jr, J. B. (1991). Synthesis of nitric oxide from L-arginine: a recently discovered pathway induced by cytokines with antitumour and antimicrobial activity. Research in immunology, 142(7), 565-569. https://doi.org/10.1016/0923-2494(91)90103-P

Hijmans, B. S., Grefhorst, A., Oosterveer, M. H., & Groen, A. K. (2014). Zonation of glucose and fatty acid metabolism in the liver: mechanism and metabolic consequences. Biochemistry, 96, 121-129. https://doi.org/10.1016/j.biochi.2013.06.007

Hmelak Gorenjak, A., & Cencič, A. (2013). Nitrate in vegetables and their impact on human health. A review. Acta alimentaria, 42(2), 158-172. https://doi.org/10.1556/aalim.42.2013.2.4

Hopkinns W. G, (2003). Plant physiology. Ed. de Boeck, Belgium, 514p.

Horner, R. F. (1982). Suspected ammonium nitrate fertilizer poisoning in cattle. The Veterinary Record, 110(20), 472-474. https://doi.org/10.1136/vr.110.20.472

Hossain, M.E., Shahrukh, S., Hossain, S.A. (2022). Chemical Fertilizers and Pesticides: Impacts on Soil Degradation, Groundwater, and Human Health in Bangladesh. In: Singh, V.P., Yadav, S., Yadav, K.K., Yadava, R.N. (eds) Environmental Degradation: Challenges and Strategies for Mitigation. Water Science and Technology Library, vol 104. Springer, Cham. https://doi.org/10.1007/978-3-030-95542-7_4

Houlot, R. (1984). Techniques d'histologie et de cytologie. Paris: Edition Maloine human disease: an overview. Methods Enzymol, 186, 1-85

Huang, J., Xu, C. C., Ridoutt, B. G., Wang, X. C., & Ren, P. A. (2017). Nitrogen and phosphorus losses and eutrophication potential associated with fertilizer application to cropland in China. Journal of Cleaner Production, 159, 171-179. https://doi.org/10.1016/j.jclepro.2017.05.008

Huang, W. H., & Lin, J. L. (2004). Acute renal failure following ingestion of manganese- containing fertilizer. *Journal of Toxicology: Clinical Toxicology*, *42*(3), 305-307. https://doi.org/10.1081/CLT-120037433.

Huang, W. H., & Lin, J. L. (2004). Acute renal failure following ingestion of manganese- containing fertilizer. Journal of Toxicology: Clinical Toxicology, 42(3), 305-307. https://doi.org/10.1081/CLT-120037433

Ihedioha, J. I., & IDIKA, I. (2007b). Contamination of drinking water with low sub-lethal levels of inorganic fertilizers: effects on body weight gain, feed and water consumption and relative organ weight percentages. Animal Production (NSAP), 32, 31-35.

Ihedioha, J. I., & Idika, I. K. (2007a). Hematological abnormalities associated with contamination of drinking water with low sub-lethal concentrations of inorganic fertilizers. In Proc 32nd Ann Conf Nig Soc Anim Prod 32: 27-30. Moller, 1995

Jublanc, C., & Bruckert, E. (2004). Hypothyroidism and cardiovascular disease: role of new risk factors and coagulation parameters. In Seminars in Vascular Medicine (Vol. 4, No. 02, pp. 145- 151). Copyright© 2004 by Thieme Medical Publishers, Inc. 333 Seventh Avenue, New York, NY 10001 USA. https://doi.org/10.1055/s-2004-835372

Justnes, H. (2002). Calcium nitrate as corrosion inhibitor for reinforced concrete. Innovations and Development in Concrete Materials and Construction, Thomas Telford Publishing, 391-401.

Justnes, H., & Nygaard, E. C. (1995). Technical calcium nitrate as set accelerator for cement at low temperatures. Cement and Concrete Research, 25(8), 1766-1774.

Kattaia, A. A., Abd El-Baset, S. A., Mohamed, E. M., Abdul-Maksou, R. S., & Elfakharany,
Y. M. (2017). Molecular mechanisms underlying histological and biochemical changes induced by nitrate in rat liver and the efficacy of S-Allylcysteine. Ultrastructural Pathology, 41(1), 10-22. https://doi.org/10.1080/01913123.2016.1252821

Kéita C, (1985). Fumures minérales et Organiques. Mémoire d'Ingénieur du Développement Rural/Option Agronomie. IDR/UO. Burkina Faso. 90p.

Kéita, C. (1985). Fumures minérales et organiques. *Mémoires d'ingénieur du Développement Rural. Option: Agronomy. IDR/UO. Burkina Faso. 90p.*

Khan, M. A., Sajid, M., Hussain, Z., Rab, A., Marwat, K. B., Wahid, F. I., & Bibi, S. (2013). How nitrogen and phosphorus influence the phenology of okra. Pakistan Journal of Botany, 45(2), 479-482.Goyal, 2012

Khan, R.A., Saxena, S.K. (1997). Integrated management of root knot nematode Meloidogyne, javanica infecting tomato using organic materials and Paecilomyces lilacininus. Biores. Technol. 61, 247-250

Killingmo, O. H., & Luthman, J. (1968). Toxicity of fertilizer grade calcium nitrate. Acta Agriculturae Scandinavica, 18(1-2), 80-86. https://doi.org/10.1080/00015126809435011

Krim, M., Messaadia, A., Maidi, I., Aouacheri, O., & Saka, S. (2013). Protective effect of ginger against toxicity induced by chromate in rats. In Annales de biologie clinique (Vol. 71, No. 2, pp. 165-173).

Kumar, R., Kumar, R., & Prakash, O. (2019). Chapter-5 the impact of chemical fertilizers on our environment and ecosystem. *Chief Ed*, *35*(69), 1173-1189.

Lavanya, K. R., Kadalli, G. G., Patil, S., Jayanthi, T., Naveen, D. V., & Channabasavegowda,

R. (2019). Sulphur Fractionation Studies in Soils of Long-Term Fertilizer Experiment under Finger Millet-Maize Cropping Sequence. International Journal of Current Microbiology and Applied Sciences, 8(9), 1334-1345.

Lazali, M., Brahimi, S., Benadis, C., & Drevon, J. J. (2020). Strategies and mechanisms of legume adaptation to low soil phosphorus availability. Revue Marocaine des Sciences Agronomiques et Vétérinaires, 8(3).

Leaf, C. D., Wishnok, J. S., & Tannenbaum, S. R. (1989). L-arginine is a precursor for nitrate biosynthesis in humans. Biochemical and biophysical research communications, 163(2), 1032- 1037. https://doi.org/10.1016/0006-291X(89)92325-5

Leclaire, A., & Monier, J. C. (1970). Structure cristalline du nitrate de calcium tetrahydrate, Comptes Rendus des Seances de l'Academie des Sciences. Serie C: Sci. Chim, 271, 1555-1557.

Lee, M. J., Chung, Y. H., Choi, H. Y., & Cha, H. G. (2020). Evaluation of subchronic repeated administration toxicity of ammonium nitrate in rats. Toxicological Research, 36(2), 115-122.

Ley, D. H. (1986). Nitrite poisoning in herring gulls (Larus argentatus) and ring-billed gulls (Larus delawarensis). Journal of wildlife diseases, 22(3), 381-384. https://doi.org/10.7589/0090-3558- 22.3.381

Li, M., Wiedmann, T., & Hadjikakou, M. (2019). Towards meaningful consumption-based planetary boundary indicators: The phosphorus exceedance footprint. Global

Environmental Change, 54, 227-238. https://doi.org/10.1016/j.gloenvcha.2018.12.005

Lindhauer, M. G. (1985). Influence of K nutrition and drought on water relations and growth of sunflower (Helianthus annuus L.). Zeitschrift für Pflanzenernährung und Bodenkunde, 148(6), 654-669. https://doi.org/10.1002/jpln.19851480608

Lindhauer, M. G. (1985). The role of potassium in the plant with emphasis on stress conditions (water, temperature, salinity). In *Proceedings of the Potassium Symposium. Pretoria, October* (pp. 95-113).

Liu, Q., Xu, H., Yi, H. (2021). Impact of Fertilizer on Crop Yield and C: N: P Stoichiometry in Arid and Semi-Arid Soil. *Int. J. Environ. Res. Public Health*, *18*, 4341. https://doi.org/10.3390/ijerph18084341.

Lockhart, K. M., King, A. M., & Harter, T. (2013). Identifying sources of groundwater nitrate contamination in a large alluvial groundwater basin with highly diversified intensive agricultural production. Journal of contaminant hydrology, 151, 140-154. https://doi.org/10.1016/j.jconhyd.2013.05.008

Luce, M. S., Whalen, J. K., Ziadi, N., & Zebarth, B. J. (2011). Nitrogen dynamics and indices to predict soil nitrogen supply in humid temperate soils. Advances in agronomy, 112, 55-102. https://doi.org/10.1016/B978-0-12-385538-1.00002-0

Lupatini, M., Korthals, G. W., De Hollander, M., Janssens, T. K., & Kuramae, E. E. (2017). Soil microbiome is more heterogeneous in organic than in conventional farming system. Frontiers in microbiology, 7, 2064. https://doi.org/10.3389/fmicb.2016.02064

Majumder, A. K., Govindarajan, B., Sharma, T., Ray, H. S., & Rao, T. C. (1995). An empirical model for chloridising-roasting of potassium in glauconitic sandstone. International journal of mineral processing, 43(1-2), 81-89. https://doi.org/10.1016/0301-7516(94)00041-W

Mallem, L., Loukil, B., & Boulakoud, M.S. (2015). The effect of fertilizers on the health of workers in the occupational environment. Ecophysiology and Toxicology, 3(1) :16-21.

Mansour, S. A., & Mossa, A. T. H. (2010). Oxidative damage, biochemical and histopathological alterations in rats exposed to chlorpyrifos and the antioxidant role of zinc. Pesticide Biochemistry and Physiology, 96(1), 14-23. https://doi.org/10.1016/j.pestbp.2009.08.008

Marschner, H. (1995). Mineral nutrition of higher plants. AcademicPress, London.

Mazzetto, A. M., Styles, D., Gibbons, J., Arndt, C., Misselbrook, T., & Chadwick, D. (2020).
Region-specific emission factors for Brazil increase the estimate of nitrous oxide emissions from nitrogen fertiliser application by 21%. Atmospheric Environment, 230, 117506. https://doi.org/10.1016/j.atmosenv.2020.117506

McLaughlin, M. J., Alston, A. M., & Martin, J. K. (1988). Phosphorus cycling in wheat-pasture rotations. III. Organic phosphorus turnover and phosphorus cycling. Soil Research, 26(2), 343-.
353. https://doi.org/10.1071/SR9880343

Mengel, K. (1997). Impact of potassium on crop yield and quality with regard to economical and ecological aspects. pp. 157-174 in Proceedings of the IPI Regional Workshop on: Food Security in the WANA Region, the Essential Need for Balanced Fertilization, Turkey International Potash Institute

Messaadia, A., Saka, S., Krim, M., Maidi, I., Aouacheri, O., & Djafer, R. (2013). Ginger- supplemented diet ameliorates ammonium nitrate-induced oxidative stress in rats. African journal of biotechnology, 12(40). https://doi.org/10.5897/AJB2013.13118

Moller, H. (1995). Adverse health effects of nitrate and its metabolites: epidemiological studies in humans. In *PotI Workshop, (Ed.), Health Aspects of Nitrates and its Metabolites (Particularly Nitrite), Council of Europe Press, Strasbourg Cedex, France* (pp. 255-268).

Moltchanova, E., Rytkönen, M., Kousa, A., Taskinen, O., Tuomilehto, J., & Karvonen for the Spat Study Group, M. (2004). Zinc and nitrate in the ground water and the incidence of Type 1 diabetes in Finland. Diabetic Medicine, 21(3), 256-261. https://doi.org/10.1111/j.1464- 5491.2004.01125.x

Moncada, S., Palmer, R. M., & Gryglewski, R. J. (1986). Mechanism of action of some inhibitors of endothelium-derived relaxing factor. Proceedings of the National Academy of Sciences, 83(23), 9164-9168. https://doi.org/10.1073/pnas.83.23.9164

Morel, C. (2002). Characterization of soil phosphorus phytoavailability by modeling phosphate ion transfer between soil and solution. Habilitation INRA Bordeaux 80p.

Morel, C., Tiessen, H. and Stewart, J. (1996). Sensitivity to equilibration periods of phosphate sorption and isotopic exchange methods assessing Q/I relationship in soils. Soil Science, 161(7): 459-467.

**Mylona, K.; Maragkoudakis, P.; Miko, L.; Bock, A.-K.; Wollgast, J.; Caldeira, S.; Ulberth,
F. (2018).** Future of food safety and nutrition-Seeking win-wins, coping with trade-offs. *Food Policy*, *74*, 143-146 https://doi.org/10.1016/j.foodpol.2017.12.002.

Nujić, M., & Habuda-Stanić, M. (2017). Nitrates and nitrites, metabolism and toxicity. Food Health Dis, 6, 63-73.

Olk, D. C., Samson, M. I., & Gapas, P. (2007). Inhibition of nitrogen mineralization in young humic fractions by anaerobic decomposition of rice crop residues. European Journal of Soil Science, 58(1), 270-281. https://doi.org/10.1111/j.13652389.2006.00836.x

WHO. (2011). World Health Organization. Nitrate and nitrite in drinking-water, Geneva, Switzerland: WHO Press, pp. 1-4.

WHO. (2006). National Health Organization. Nitrate and potential endogenous formation of N- nitroso compounds Food Additives Series: no 50 Who. Geneva, P1-9 24

Oram, B. (2015). Nitrates and Nitrites in Drinking Water and Surface Waters. Water Research Center. http://www.waterresearch.net/index.php/nitrate [6.2.2018].

Organization for Economic Cooperation and Development (OECD). (2000). Guidance Notes for the Analysis and Evaluation of Repeat Dose Toxicity Studies. Environment, Health and Safety Publications Series on Testing and Assessment, Environment Directorate, OECD, Paris, pp. 27- 72.

Osunlaja, S.O. (1989). Effect of organic soil amendments on the incidence of brown spot disease in maize caused by Physoderma maydis. J. Basic Microbiol. 29: 501- 505

Pachauri, R. K., & Meyer, L. A. (2014). Climate Change 2014: Synthesis Report. Contribution of Working Groups I, II and III to the Fifth Assessment Report of the Intergovernmental Panel on Climate Change. IPCC, Geneva, Switzerland, 151 pp.

Pahalvi, H.N., Rafiya, L., Rashid, S., Nisar, B., Kamili, A.N. (2021). Chemical Fertilizers and Their Impact on Soil Health. In: Dar, G.H., Bhat, R.A., Mehmood, M.A., Hakeem, K.R. (eds) Microbiota and Biofertilizers, Vol 2. Springer, Cham.

https://doi.org/10.1007/978-3-030-61010-4_1

Pearce, E. N. (2004). Hypothyroidism and dyslipidemia: modern concepts and approaches. Current cardiology reports, 6(6), 451-456. https://doi.org/10.1007/s11886-004-0054-3

Pereda Campos, M. V. (2008). Contribution to the study of phosphate transporters of the PHT1 family in Poplar (Populus trichocarpa Torr. & Gray) and the ectomycorrhizal fungus Laccaria bicolor (Maire) P. D. Orton. Doctoral thesis. Université Henri Poincaré, Nancy I. Spécialité : Biologie Végétale et Forestière. 140 p

Pereira, C., Ferreira, N. R., Rocha, B. S., Barbosa, R. M., & Laranjinha, J. (2013). The redox interplay between nitrite and nitric oxide: From the gut to the brain. Redox, 1(1), 276-284. https://doi.org/10.1016/j.redox.2013.04.004

Pierre, W. H. (1928). Nitrogenous fertilizers and soil acidity. I. Effect of various nitrogenous fertilizers on soil reaction.

Pierzynski, G. M., McDowell, R. W., & Thomas Sims, J. (2005). Chemistry, cycling, and potential movement of inorganic phosphorus in soils. *Phosphorus: Agriculture and the environment*, *46*, 51-86. https://doi.org/10.2134/agronmonogr46.c3

Pierzynski, G.M., Sims, J.T. & Vance, G.F. (2000). Soils and Environmental Quality, 2nd ed. CRC Press, Boca Raton, FL. 459 p

Portmann, R. W., Daniel, J. S., & Ravishankara, A. R. (2012). Stratospheric ozone depletion due to nitrous oxide: influences of other gases. Philosophical Transactions of the Royal Society B: Biological Sciences, 367(1593), 1256-1264. https://doi.org/10.1098/rstb.2011.0377

Postma-Blaauw, M. B., Bloem, J., Faber, J. H., Van Groenigen, J. W., De Goede, R. G., & Brussaard, L. (2006). Earthworm species composition affects the soil bacterial community and net nitrogen mineralization. Pedobiologia, 50(3): 243-256. https://doi.org/10.1016/j.pedobi.2006.02.001

Priya P., SAHI S.V. (2009). Inflfluence of phosphorus nutrition on growth and metabolism of Duo grass (Duo festulolium). Plant Physiology and Biochemistry, 47 :31-36.

Radomski, M. W., Palmer, R. M., & Moncada, S. (1987). The anti-aggregating properties of vascular endothelium: interactions between prostacyclin and nitric oxide. British journal of pharmacology, 92(3), 639. https://doi.org/10.1111%2Fj.1476-5381.1987.tb11367.x

Raidal, S. R., & Jaensch, S. M. (2006). Acute poisoning of silver gulls (Larus novaehollandiae) following urea fertilizer spillage. Avian pathology, 35(1), 38-41. https://doi.org/10.1080/03079450500465718

Ramade, F. (1998). Encyclopedic dictionary of water sciences (No. 631.4 R165d). Paris, FR: Ediscience, 1998.

Rastogi, R. P., Madamwar, D., & Incharoensakdi, A. (2015). Bloom dynamics of cyanobacteria and their toxins: environmental health impacts and mitigation strategies. Frontiers in microbiology, 6, 1254. https://doi.org/10.3389/fmicb.2015.01254

Rodriguez, M., & Zea, H. (2015). Evaluation of a synthesis process for the production of calcium nitrate liquid fertilizer. *International Journal of ChemTech Research,* 7(4):1960-1965. http://sphinxsai.com/2015/ch_vol7_no4/5/(1960-1965)%20V7N4.pdf

Ropp, R. C. (2013). Chapter4 Group 15 (N, P, Sb and Bi) alkaline earth compounds. In:R.C. Ropp (Ed.), Elsevier, Amsterdam, Encyclopedia of the alkaline earth compounds, 199-350.

Rouag, M., Berrouague, S., Djaber, N., Khaldi, T., Boumendjel, M., Taibi, F., ... & Messarah,
M. (2020). Pumpkin seed oil alleviates oxidative stress and liver damage induced by sodium nitrate in adult rats: biochemical and histological approach. African Health Sciences, 20(1), 413-425. https://doi.org/10.4314/ahs.v20i1.48

Saggar, S., Jha, N., Deslippe, J., Bolan, N. S., Luo, J., Giltrap, D. L., ... & Tillman, R. W. (2013). Denitrification and N2O: N2 production in temperate grasslands: Processes, measurements, modelling and mitigating negative impacts. Science of the Total

Environment, 465: 173-195.

Sanchez E., Muñoz E., Anchondo Á., Ruiz J.M., Romero L., (2009). Nitrogen impact on nutritional status of Phosphorus and its main bioindicator: response in the roots and leaves of green bean plants. Revista Chapingo. Serie horticultura, 15(2): 177-182.

Sanginga, N., & Woomer, P. L. (Eds.). (2009). *Integrated soil fertility management in Africa: principles, practices, and developmental process*. CIAT.

Sato, A., Gonmori, K., & Yoshioka, N. (1999). A case of fatal intoxication with ammonium sulfate and a toxicological study using rabbits. Forensic science international, 101(2), 141- 149. https://doi.org/10.1016/S0379-0738(99)00014-6

Sauchelli, V. (1960). Chemistry and Techno/ogy of Ferlilizers. Reinhold Publishing Corporation. New York.

Schvartz, C., J. Decroux & J.C. Muller. Muller (2005). Guide de la fertilisation raisonnée: grandes cultures et prairies. Comité français d'étude et de développement de la fertilisation raisonnée. France Agricole Editions, 2005. 414 p.

Sellers, R. S., Mortan, D., Michael, B., Roome, N., Johnson, J. K., Yano, B. L., ... & Schafer, K. (2007). Society of Toxicologic Pathology position paper: organ weight recommendations for toxicology studies. Toxicologic pathology, 35(5), 751-755. https://doi.org/10.1080/01926230701595300

Senesi, N. (1989). Composted materials as organic fertilizers. *Science of the Total Environment*, *81*, 521-542. https://doi.org/10.1016/0048-9697(89)90161-7

Shah, F., & Wu, W. (2019). Soil and crop management strategies to ensure higher crop productivity within sustainable environments. Sustainability, 11(5), 1485. https://doi.org/10.3390/su11051485

Sharma, B., & Ahlert, R. C. (1977). Nitrification and nitrogen removal. Water Research, 11(10), 897-925. https://doi.org/10.1016/0043-1354(77)90078-1

Sharma, N., & Singhvi, R. (2017). Effects of chemical fertilizers and pesticides on human health and environment: a review. *International journal of agriculture, environment and biotechnology*, *10*(6), 675-680.

Sheahan, M., Barrett, C. B., & Goldvale, C. (2017). Human health and pesticide use in sub-Saharan Africa. Agricultural Economics, 48(S1), 27-41. https://doi.org/10.1111/agec.12384

Sofilić, T. (2014). Ekotoksikologija. University of Zagreb, Faculty of Metallurgy, Sisak, Croatia,
pp. 24-25.

Steffen, W., Richardson, K., Rockström, J., Cornell, S. E., Fetzer, I., Bennett, E. M., ... & Sörlin, S. (2015). Planetary boundaries: Guiding human development on a changing planet. science, 347(6223), 1259855. https://doi.org/10.1126/science.1259855

Stevenson, F. J. (1986). The internal cycle of nitrogen of soil: carbon, nitrogen, phosphorus, sulfur, micronutrients. John Wiley & Sons, Inc, New York, NY. 380 p.

Strous, M., Kuenen, J. G., & Jetten, M. S. (1999). Key physiology of anaerobic ammonium oxidation. Applied and environmental microbiology, 65(7), 3248-3250. https://doi.org/10.1128/AEM.65.7.3248-3250.1999

Sullivan, B. W., Smith, W. K., Townsend, A. R., Nasto, M. K., Reed, S. C., Chazdon, R. L., & Cleveland, C. C. (2014). Spatially robust estimates of biological nitrogen (N) fixation imply substantial human alteration of the tropical N cycle. Proceedings of the National Academy of Sciences, 111(22), 8101-8106.

Talboys, P. J., Heppell, J., Roose, T., Healey, J. R., Jones, D. L., & Withers, P. J. (2016). Struvite: a slow-release fertiliser for sustainable phosphorus management? Plant and soil, 401(1), 109-123. https://doi.org/10.1007/s11104-015 2747-3

Taliman, N. A., Dong, Q., Echigo, K., Raboy, V., & Saneoka, H. (2019). Effect of phosphorus fertilization on the growth, photosynthesis, nitrogen fixation, mineral accumulation, seed yield, and seed quality of a soybean low-phytate line. Plants, 8(5), 119. https://doi.org/10.3390/plants8050119

Testud, F. (2004). Mineral fertilizers. EMC-Toxicologie-Pathologie, 1(1), 21-28.

Testud, F. (2015). Mineral fertilizers. EMC (Elsevier-Masson SAS, Paris), Pathologie

Professionnelle et de l'Environnement, 10(3). https://.doi.org/10.1016/S1877-7856(15)62787-8

Tian, H., Yang, J., Xu, R., Lu, C., Canadell, J. G., Davidson, E. A., ... & Zhang, B. (2019). Global soil nitrous oxide emissions since the preindustrial era estimated by an ensemble of terrestrial biosphere models: Magnitude, attribution, and uncertainty. Global change biology, 25(2), 640-659. https://doi.org/10.1111/gcb.14514

Tootian, Z., Monfared, A. L., Fazelipour, S., Shybani, M. T., Rouhollah, F., Sasani, F.& Molaemi, E. (2012). Biochemical and structural changes of the kidney in mice exposed to phenol. Turkish Journal of Medical Sciences, 42(4), 695-703. https://doi.org/10.3906/sag-1012-1332.

Trenkel, ME.T. (2021). Slow-and Controlled-Release and Stabilized Fertilizers: An Option for Enhancing Nutrient Use Effiiency in Agriculture. International Fertilizer Industry Association (IFA).

Valadares, R. V., Cantarutti, R. B., Mattiello, E. M., & Vieira, R. F. (2017). Agronomic effectiveness of rock phosphate combined with nitrogen sources in spot application: A pot experiment. Journal of Plant Nutrition and Soil Science, 180(5), 585-593. https://doi.org/10.1002/jpln.201600037

Vitousek, P. M., Naylor, R., Crews, T., David, M. B., Drinkwater, L. E., Holland, E., & Zhang,
F. S. (2009). Nutrient imbalances in agricultural development. Science, 324(5934), 1519-1520. https://doi.org/10.1126/science.1170261

Volkmer, B. G., Ernst, B., Simon, J., Kuefer, R., Bartsch Jr, G., Bach, D., & Gschwend, J. E. (2005). Influence of nitrate levels in drinking water on urological malignancies: a community- based cohort study. BJU international, 95(7), 972-976. https://doi.org/10.1111/j.1464- 410X.2005.05450.x

Wagner, D. A., Schultz, D. S., Deen, W. M., Young, V. R., & Tannenbaum, S. R. (1983). Metabolic fate of an oral dose of 15 N-labeled nitrate in humans: effect of diet supplementation with ascorbic acid. Cancer research, 43(4), 1921-1925.

Wang, Y., Ying, H., Yin, Y., Zheng, H., & Cui, Z. (2019). Estimating soil nitrate leaching of nitrogen fertilizer from global meta-analysis. Science of the Total

Environment, 657, 96-102. https://doi.org/10.1016/j.scitotenv.2018.12.029

Wang, Z.; Hassan, M.U.; Nadeem, F.; Wu, L.; Zhang, F.; Li, X. (2020). Magnesium Fertilization Improves Crop Yield in Most Production Systems: A Meta-Analysis. *Front. Plant Sci. 10*, 1727. https://doi.org/10.3389/fpls.2019.01727.

Weil, C. S., & McCollister, D. D. (1963). Safety evaluation of chemicals, relationship between short-and long-term feeding studies in designing an effective toxicity test. Journal of Agricultural and Food Chemistry, 11(6), 486-491.

Weissenberger, J., Duffourg, J. M., Hepner, S., & Morano, D. (2006, January). Reducing odors from biosolids with the help of a nitrate salt. In Residuals and Biosolids Conference 2006 (pp. 109- 119). Water Environment Federation. https://doi.org/10.2175/193864706783797276.

WHO. (2006). Organization national de santé. Nitrate and potential endogenous formation of N- nitroso combounds Food Additives Series: no 50 Who. Geneva, P1-9 24

Wrathall, C. (2002): Nitrate toxicity, Indiana Animal Disease Diagnostic Laboratory https://www.addl.purdue.edu/newsletters/2002/fall/ fall2002.pdf [6.2.2018].

Yadav, R. L., Dwivedi, B. S., Prasad, K., Tomar, O. K., Shurpali, N. J., & Pandey, P. S. (2000). yield trends, and changes in soil organic-C and available NPK in a long-term rice-wheat system under integrated use of manures and fertilizers. Field Crops Research, 68(3), 219-246. https://doi.org/10.1016/S0378-4290(00)00126-X

Yeruham, I., Shlosberg, A., Hanji, V., Bellaiche, M., Marcus, M., & Liberboim, M. (1997). Nitrate toxicosis in beef and dairy cattle herds due to contamination of drinking water and whey. Veterinary and human toxicology, 39(5), 296-298.

Yousaf, M., Li, J., Lu, J., Ren, T., Cong, R., Fahad, S., & Li, X. (2017). Effects of fertilization on crop production and nutrient-supplying capacity under rice-oilseed rape rotation system. Scientific reports, 7(1), 1270.https://doi.org/10.1038/s41598-017-01412-0

Yuniwati, E. D., Darmayanti, R., & Farooq, S. M. Y. (2023). How is organic fertilizer produced and applied to chili and eggplant plants. AMCA Journal of Community Development, 3(2), 88-94. https://doi.org/10.51773/ajcd.v3i2.300

Zhong, Y., Yan, W., & Shangguan, Z. (2015). Impact of long-term N additions upon coupling between soil microbial community structure and activity, and nutrient-use efficiencies. Soil Biology and Biochemistry, 91, 151-159. https://doi.org/10.1016/j.soilbio.2015.08.030

Zhou, J., Guan, D., Zhou, B., Zhao, B., Ma, M., Qin, J., ... & Li, J. (2015). Influence of 34- years of fertilization on bacterial communities in an intensively cultivated black soil in northeast China. Soil Biology and Biochemistry, 90, 42-51. https://doi.org/10.1016/j.soilbio.2015.07.005

More Books!

I want morebooks!

Buy your books fast and straightforward online - at one of world's fastest growing online book stores! Environmentally sound due to Print-on-Demand technologies.

Buy your books online at
www.morebooks.shop

Kaufen Sie Ihre Bücher schnell und unkompliziert online – auf einer der am schnellsten wachsenden Buchhandelsplattformen weltweit! Dank Print-On-Demand umwelt- und ressourcenschonend produzi ert.

Bücher schneller online kaufen
www.morebooks.shop

info@omniscriptum.com
www.omniscriptum.com

OMNIScriptum

MIX
Papier aus verantwortungsvollen Quellen
Paper from responsible sources
FSC® C105338

Printed by Books on Demand GmbH, Norderstedt / Germany